Standard Grade | General

Biology

General Level 2001

General Level 2002

General Level 2003

General Level 2004

General Level 2005

First exam published in 2001.

Published by Leckie & Leckie, 8 Whitehill Terrace, St. Andrews, Scotland KY16 8RN tel: 01334 475656 fax: 01334 477392

enquiries@leckieandleckie.co.uk www.leckieandleckie.co.uk

ISBN 1-84372-289-5

A CIP Catalogue record for this book is available from the British Library.

Printed in Scotland by Scotprint.

Leckie & Leckie is a division of Granada Learning Limited, part of ITV plc.

Acknowledgements

Leckie & Leckie is grateful to the copyright holders, as credited at the back of the book, for permission to use their material.

Every effort has been made to trace the copyright holders and to obtain their permission to use their copyright material.

Leckie & Leckie will gladly receive information enabling them to rectify any error or omission in subsequent editions.

[BLANK PAGE]

FOR OFFICIAL USE

G

KU   PS

Total Marks

0300/401

NATIONAL
QUALIFICATIONS
2001

MONDAY, 21 MAY
9.00 AM – 10.30 AM

BIOLOGY
STANDARD GRADE
General Level

Fill in these boxes and read what is printed below.

Full name of centre

Town

Forename(s)

Surname

Date of birth
Day Month Year   Scottish candidate number   Number of seat

1 All questions should be attempted.

2 The questions may be answered in any order but all answers are to be written in the spaces provided in this answer book, and must be written clearly and legibly in ink.

3 Rough work, if any should be necessary, as well as the fair copy, is to be written in this book. Additional spaces for answers and for rough work will be found at the end of the book. Rough work should be scored through when the fair copy has been written.

4 Before leaving the examination room you must give this book to the invigilator. If you do not, you may lose all the marks for this paper.

SCOTTISH
QUALIFICATIONS
AUTHORITY
©

1. The diagram below represents part of a food web involving wheat.

Marks | KU | PS

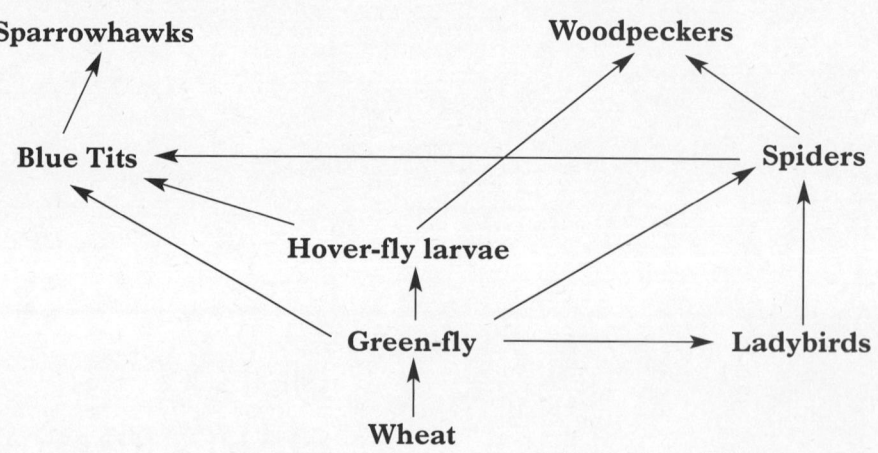

(a) Sparrowhawks are consumers.
Explain what is meant by the term consumers.

_____

_____

1

(b) How is the transfer of energy represented in a food web diagram?

_____

1

(c) Most of the energy taken in by the blue tits does not pass to the sparrowhawks which eat them.
Give **two** ways in which this energy may be lost.

1 _____

2 _____

1

Page two

Marks | KU | PS

2. The table below contains examples of pollution of four different ecosystems.

Ecosystem affected	Source of pollution	Example of pollutant
Air	Domestic	CFC gases from aerosol sprays
Fresh water		Pesticides in a river
	Industrial	Crude oil from tanker vessels
Land	Domestic	

(a) Complete the empty boxes in the table.

3

(b) The list below contains statements about pollution.

    X   Smoke from coal fired power stations causes acid rain.

    Y   Raw sewage in rivers leads to the death of fish.

    Z   Car exhaust fumes contain poisonous gases.

Choose **one** of the statements and give an example of a way in which the pollution **could** be controlled.

Statement letter _____

Method of control _____

_____

1

[Turn over

3. The table below shows the annual percentage yield loss of five crops, due to disease and insects.

Crop plant	Percentage yield loss	
	Insects	Disease
Wheat	4	9
Rice	28	8
Barley	5	5
Oats	5	10
Maize	10	10

(a) Use the table to complete the chart below by

(i) labelling the Y-axis

(ii) adding the scale to the Y-axis

(iii) completing the bars for the other crops.

(Additional graph paper, if required, will be found on page 29.)

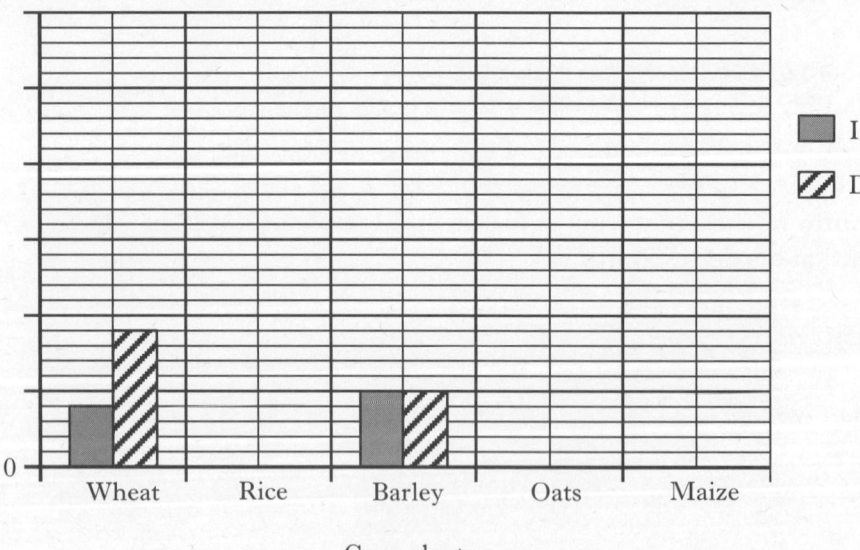

Crop plants

(b) Which crop plant had the lowest total yield loss?

_____

(c) Which crop plants have a greater yield loss to disease than to insects?

_____

DO NOT WRITE IN THIS MARGIN

Marks | KU | PS

4. (*a*) The bar chart below shows the proportion of different groups of animals and plants found in Scotland.

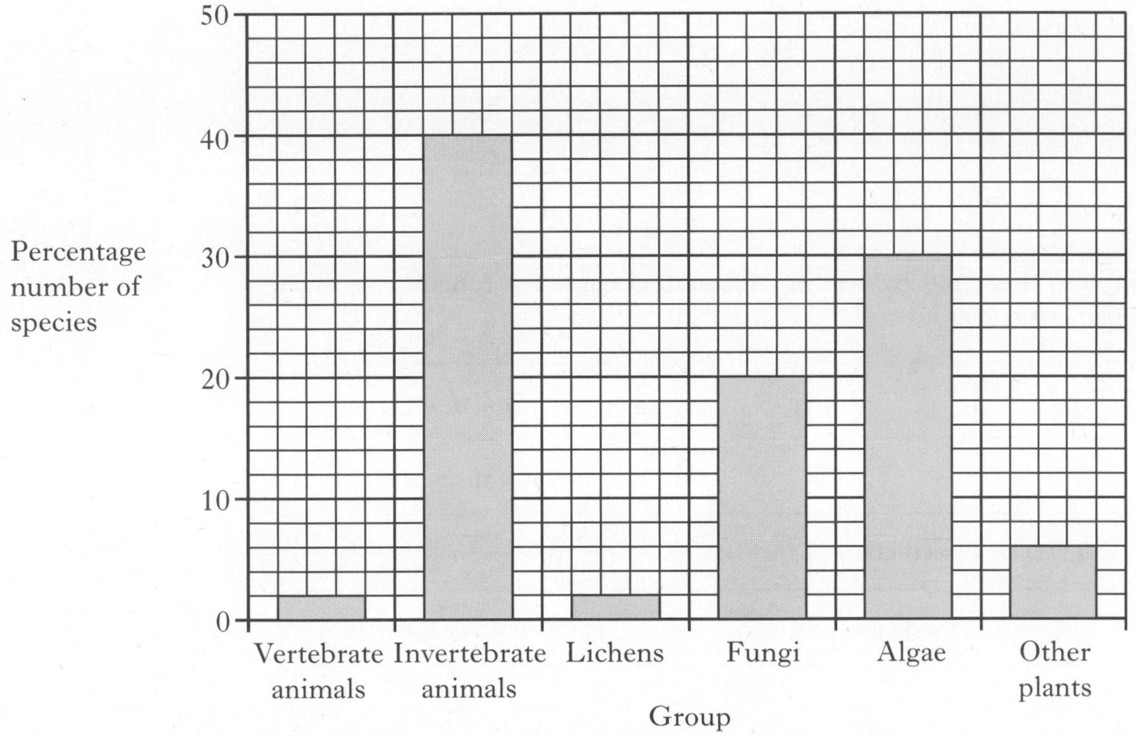

(i) What percentage of the total number of species are fungi?

_________ % **1**

(ii) What percentage of the total number of species are animals?

Space for calculation

_________ % **1**

(iii) The total number of species is 70 000. How many of these are algae?

Space for calculation

Number of algae species _________ **1**

(*b*) Many species of plants and animals are useful to humans.

Give **two** different uses of plants by humans.

1 ______________________________

2 ______________________________ **1**

[Turn over

Marks | KU | PS

5. (*a*) The diagram below shows a section through a seed.

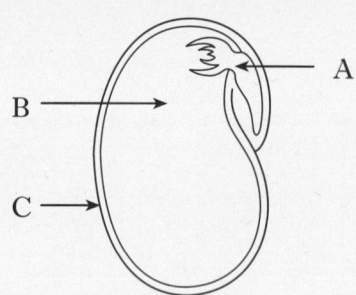

Complete the table by writing the letter, name or function of each labelled structure.

Letter	Name	Function
		forms young plant
C	seed coat	
	food store	resources for growth

2

(*b*) Gardeners can buy plant seeds from catalogues, which give information as shown in the table below.

Plant name	Flowering ability	Temperature range for germination (°C)
Busy Lizzie	◗	19 – 25
Dahlia	○	15 – 20
Marigold	○	20 – 25
Geranium	◗	20 – 24
Pansy	○	16 – 21
Dianthus	◗	20 – 25

○ = requires plenty of light for flowering   ◗ = flowers well in shade

(i) What temperature would be suitable for germination of all these seeds?

_____ °C

1

5. (*b*) **(continued)**

(ii) The chart shows the germination temperatures for the three types of plants that require plenty of light for flowering.

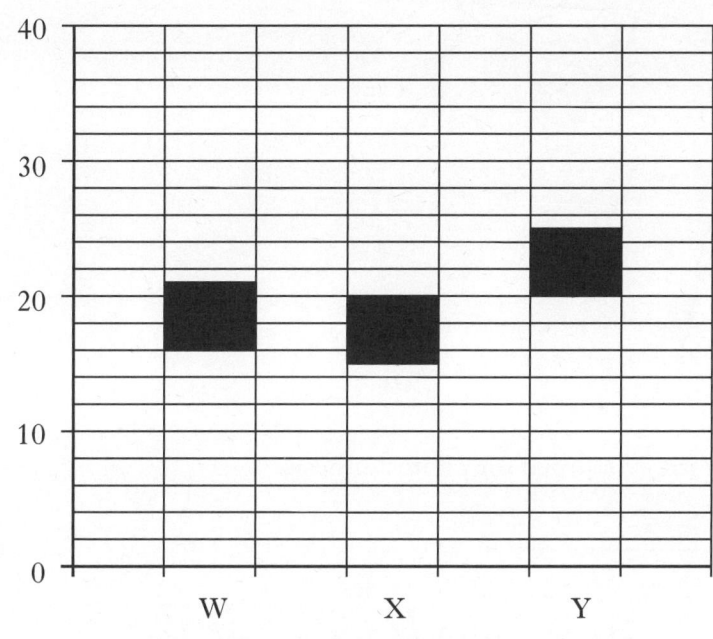

Identify plants W, X and Y.

W _____

X _____

Y _____

1

Page seven

[Turn over

Marks | KU | PS

6. (a) The diagram represents a plant carrying out photosynthesis.

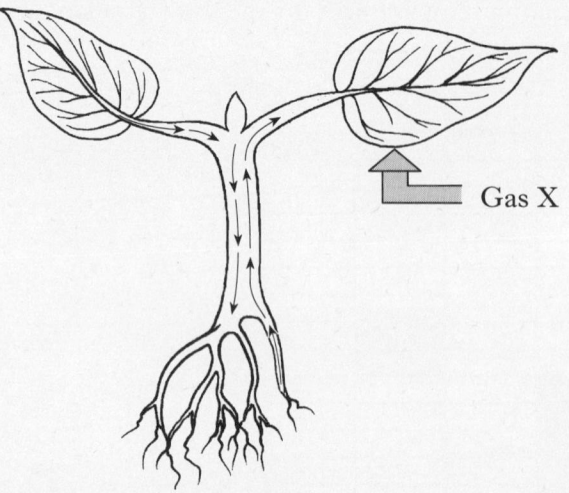

Gas X

    (i) Name Gas X which is required for photosynthesis.

    _____

                                              1

    (ii) Name the pores on the leaves, through which Gas X can enter.

    _____

                                              1

(b) Name the gas produced during photosynthesis.

_____

                                              1

(c) Name the chemical, made from glucose, which is stored in the leaves.

_____

                                              1

(d) The transport of substances in the plant is shown by arrows (↓↑) in the diagram.
Complete the table below with the correct information.

Transport tissue	Substance carried	Part of plant from which substance is carried
		root
	sugar	leaves

                                              2

Marks | KU | PS

7. The diagram below shows a cell from the leaf of a green plant.

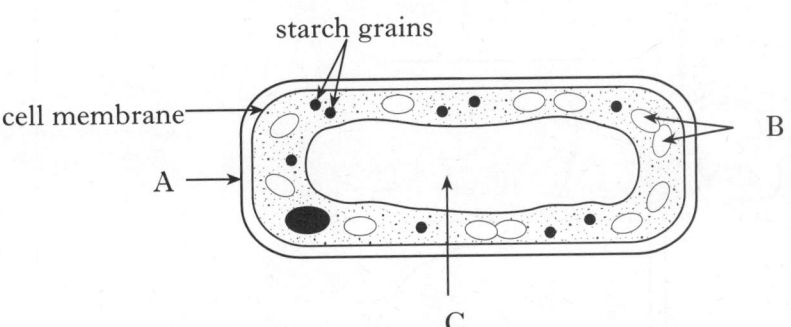

(a) Complete the table with the names of the parts shown in the diagram.

Letter	Cell part
A	
B	
C	

1

(b) Name the type of cell division which increases the number of cells for the growth of an organism.

_____

1

(c) State **one** reason why cells need energy, other than for cell division.

_____

1

(d) Complete the word equation for aerobic respiration.

Glucose + _____ ⟶ _____ + _____ + energy   1

[Turn over

8. (*a*) The diagram below shows part of the human digestive system.

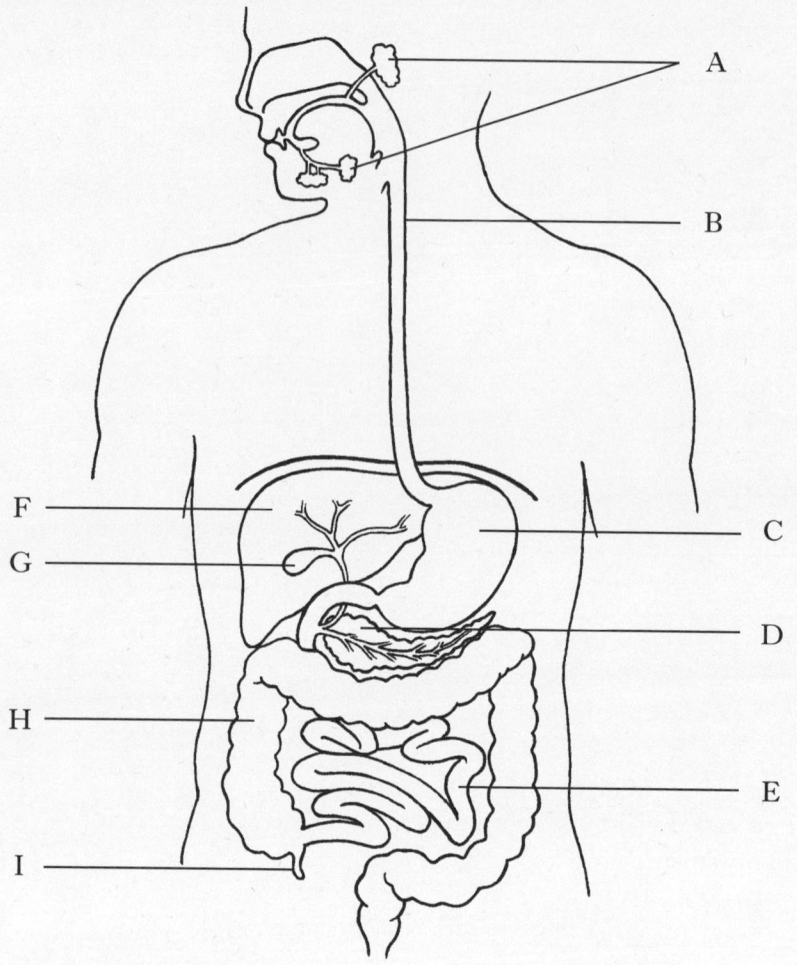

(i) Complete the following table to identify the parts it contains.

Letter	Name of part
A	
	oesophagus
D	
	liver
I	

3

(ii) The large intestine (H) eliminates undigested food from the body as faeces.

State **one other** function it performs.

_____

_____

1

Marks | KU | PS

8. (a) (continued)

(iii) Explain how each of the following features of the small intestine helps it to function efficiently.

1. The small intestine is long.

Explanation _____

_____ **1**

2. The small intestine contains many blood vessels in its walls.

Explanation _____

_____ **1**

(b) The grid below contains words about the kidneys.

renal artery	renal vein	urea
filtration	ureter	glucose
bladder	reabsorption	urine

Use words from the grid to complete the following sentences correctly.

(i) The _____ brings blood to the kidney and the

_____ takes blood away from the kidney. **1**

(ii) The kidneys are the main organs for regulating the water content of mammals.

Their method of action involves _____ of

blood followed by _____ of useful substances. **1**

(iii) _____ is a waste product which is removed

in the _____ . **1**

[Turn over

Marks | KU | PS

9. Cola is a type of fizzy drink. An investigation into its effect on teeth was carried out as shown in the diagram below.

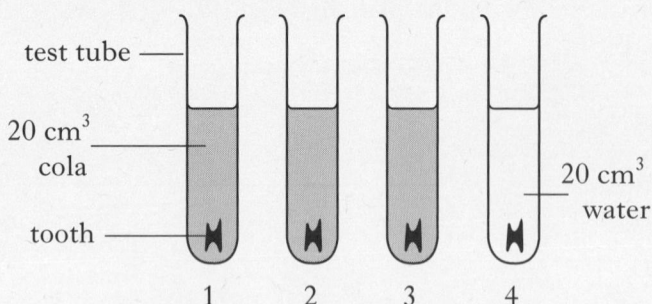

test tube

20 cm^3 cola

tooth

20 cm^3 water

1   2   3   4

(a) Complete the following table by

   (i)   adding the correct headings   **1**

   (ii)   calculating the missing percentage   **1**

   (iii)   completing the results for tooth 2.   **1**

Tooth number			Loss in weight (mg)	Percentage loss in weight
1	3000	2100	900	
2	4200			10
3	3800	3040	760	20
4 (control)	4000	4000	0	0

Space for calculations

(b) Tooth 4 was used as a control.
What is the purpose of a control?

_____

_____   **1**

(c) The teeth were sterilised before carrying out this investigation.
Explain why this was necessary.

_____

_____   **1**

DO NOT WRITE IN THIS MARGIN

Marks KU PS

9. (continued)

(*d*) Give two factors, not mentioned already, which would need to be kept constant for the investigation to be valid.

1 ____

2 ____ **2**

(*e*) What valid conclusion could be drawn from the results of the investigation?

____

____ **1**

[Turn over

Marks KU PS

10. To investigate the effect of temperature on the activity of the enzyme pepsin, five test tubes were set up as shown below.

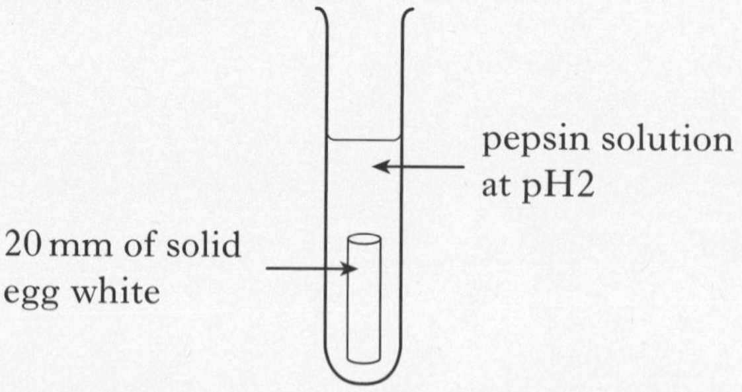

pepsin solution at pH2

20 mm of solid egg white

Each tube was placed in a water bath at a different temperature. After 12 hours, the following results were obtained.

Test tube	Temperature (°C)	Length of egg white after 12 hours (mm)
A	5	19
B	20	17
C	35	13
D	45	15
E	60	20

(a) At which temperature did the greatest digestion of egg white take place?

_____ °C

1

(b) Describe the effect of increasing the temperature on the activity of the pepsin over each of the temperature ranges below.

Between 5 °C and 35 °C _____

1

Between 35 °C and 60 °C _____

1

(c) If the experiment had been repeated at pH7, which of the following would be the most likely result for the length of egg white in test tube B?

Tick the correct box.

19 mm ☐   17 mm ☐

15 mm ☐   13 mm ☐

1

11. The diagram below represents part of the human ear.

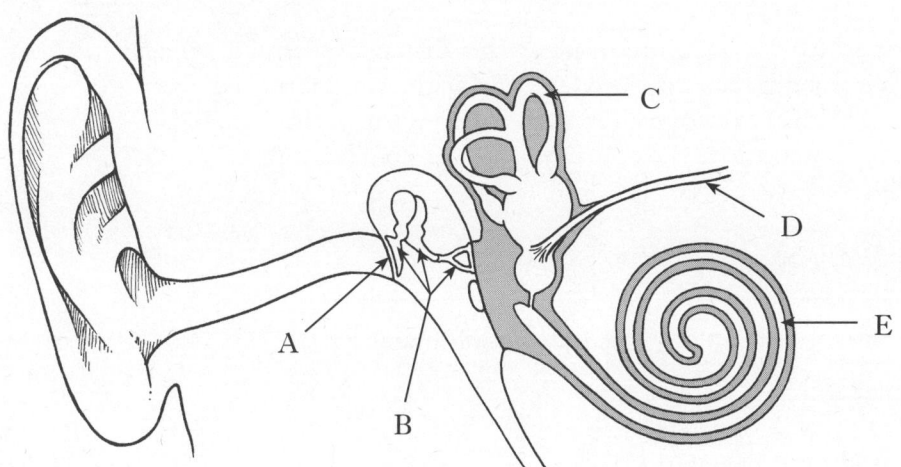

(a) Complete the table below.

Letter	Name	Function
		picks up vibrations in the air
	bones of the middle ear	amplify vibrations and pass them to the inner ear
E		changes vibrations into nerve impulses
D	auditory nerve	
C	semi-circular canals	

3

(b) What can you judge more accurately when using two ears, rather than one?

_____

1

(c) The list below gives the names of some parts of the human body.
Underline the **three** parts, which make up the nervous system.

head   heart   nerves   muscle

skin   spinal cord   lungs   brain

1

[Turn over

	Marks	KU	PS

12. (*a*) In an investigation to measure fitness, the distance sprinted by an athlete in five seconds was measured. The sprints were repeated every 15 seconds. The distance covered in each sprint is shown in the table.

Time at start of sprint (s)	0	15	30	45	60	75	90
Distance covered (m)	40	40	39	36	32	27	21

    (i) Use the table to complete the **line graph** below by

       1. labelling the X-axis   **1**

       2. adding a scale to the Y-axis   **1**

       3. completing the graph.   **1**

    Two points have already been plotted.

    (Additional graph paper, if required, will be found on page 29.)

Distance covered (m)

0   15   30   45   60   75   90

    (ii) Between which two times was there the biggest decrease in distance covered in the sprints?

       Between _____ s and _____ s   **1**

    (iii) What valid conclusion could be drawn about the distance covered in a sprint as the number of sprints increased?

    _____   **1**

    (iv) What could have been done to check that these results are reliable?

    _____   **1**

DO NOT WRITE IN THIS MARGIN

	Marks	KU	PS

12. (continued)

(*b*) Complete the following sentence by adding the names of the missing chemicals.

Muscle fatigue is caused by the lack of ____________________ and the

build up of ____________________ in muscles. **2**

[Turn over

13. Many birds feed and roost at airports. Collisions between birds and planes may result in crashes. Scientists try to use their understanding of bird behaviour to reduce the number of collisions.

The pie charts show the number of collisions with different birds at five airports.

Chart A 1994–1996　　　　　　**Chart B 1997–1999**

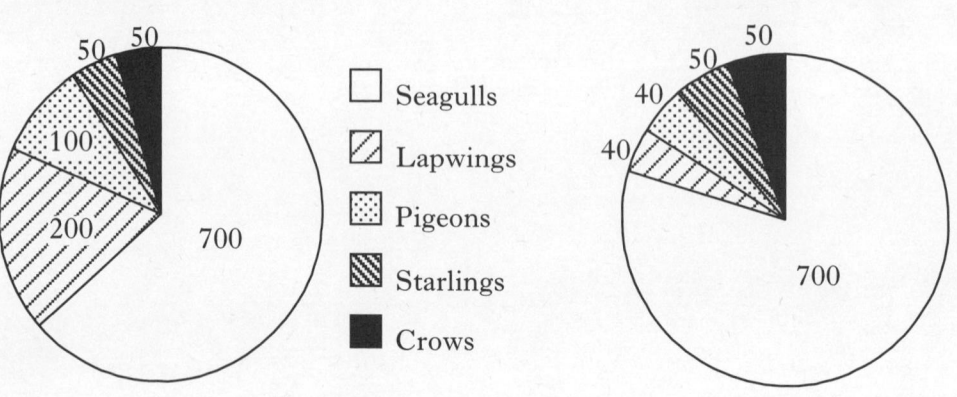

□ Seagulls

▨ Lapwings

▨ Pigeons

▨ Starlings

■ Crows

(a)　(i)　Which type of bird was involved in most collisions during the period 1994–1996?

_____

1

(ii)　What was the total number of collisions in this period?

Space for calculation

_____

1

(b)　From 1997, birds of prey were kept at these airfields.

(i)　Which **two** species were involved in fewer collisions after the introduction of the birds of prey?

_____ and _____

1

(ii)　What appeared to be the effect of the birds of prey on the number of collisions with seagulls?

_____

1

(iii)　Calculate the ratio of collisions involving lapwings before and after the introduction of the birds of prey.

Space for calculation

_____ : _____

before　after

1

Marks | KU | PS

14. (*a*) <u>Underline</u> **one** word in each group to make the sentences correct.

Yeast is a $\left\{ \begin{array}{c} bacterium \\ fungus \end{array} \right\}$ and is $\left\{ \begin{array}{c} single\text{-}celled \\ multicellular \end{array} \right\}$.

Yeast can use $\left\{ \begin{array}{c} sugar \\ oxygen \end{array} \right\}$ as a source of food.

1

(*b*) Complete the table by writing the correct word from the list in the empty boxes.

Each word may be used **once, more than once** or **not at all.**

Description	Word
Organisms used to make yoghurt	
Pieces of these can be transferred from a different organism into bacteria by genetic engineering to make new substances	
Chemicals made by micro-organisms and which kill bacteria	

List

antibiotics
bacteria
fungus
enzymes
hormones
chromosomes

3

[Turn over

Marks | KU | PS

15. Read the following passage carefully.

Salt sellers threaten the whale. Adapted from "The Sunday Herald".

Every year grey whales leave the seas around Alaska and travel to the San Ignacio Lagoon in Mexico where they mate. They return a year later to give birth and mate again. The red mangroves that edge the lagoon provide shelter for the newborn calves. The high salt content of the water gives the calves support while they learn to swim.

The whales' breeding grounds are now in danger from a Japanese/Mexican company which has applied to build a salt production plant in San Ignacio. The new plant would produce seven million tonnes of salt a year, the quantity Japan imports from Australia. Japan uses the salt in everything from the manufacture of glass to cosmetics.

Environmental groups claim that the company's manufacturing methods will alter the salt concentration of the lagoon. The methods involve removing the salt from 6600 gallons of water per second and pumping salt-free water back into the lagoon. Controversy surrounds the company's existing salt production plant nearby where the dead bodies of marine animals such as whales, turtles and fish have been found washed up on the shore. The company said the most likely explanation for this was that the animals had been killed by a chemical dye released into the water by drug traffickers.

Answer the following questions based on the passage.

(*a*) Which area do the grey whales leave to go to their breeding grounds?

_____ **1**

(*b*) State **two** reasons why the lagoon provides ideal conditions for the whale calves.

1 _____

2 _____ **1**

(*c*) How much salt does Japan import from Australia every year?

_____ tonnes **1**

Marks | KU | PS

15. **(continued)**

(d) Name **two** products, mentioned in the passage, which require salt during their manufacture.

1 _____

2 _____

1

(e) Which part of the manufacturing process affects the salt concentration of the lagoon?

_____

_____

1

(f) Does the salt production company accept responsibility for the death of marine animals in their area? Give a reason for your answer.

Accept responsibility _____

Reason for answer   _____

_____

1

[Turn over

Official SQA Past Papers: General Biology 2001

DO NOT
WRITE IN
THIS
MARGIN

Marks | KU | PS

16. The diagram below shows the sex chromosomes present in the cells of two generations.

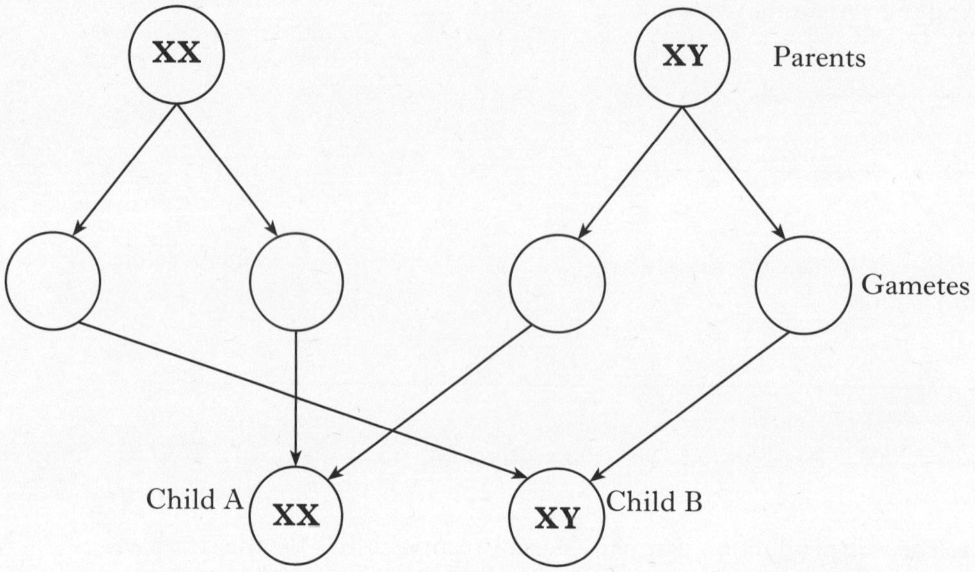

(a) (i) Complete the diagram to show the sex chromosomes of the gametes and the children.   1

   (ii) What is the sex of the children?

   Child A _____ Child B_____   1

   (iii) What symbol may be used to represent the children's generation?

   _____   1

(b) For each of the following, write the word described by the phrase.

   (i) The genes that an organism contains.

   Word _____   1

   (ii) The cell which carries one of the two forms of a gene.

   Word _____   1

   (iii) Differences between organisms of the same species.

   Word _____   1

Marks | KU | PS

17. The diagram below shows the results of an experiment using yeast and sugar solution.

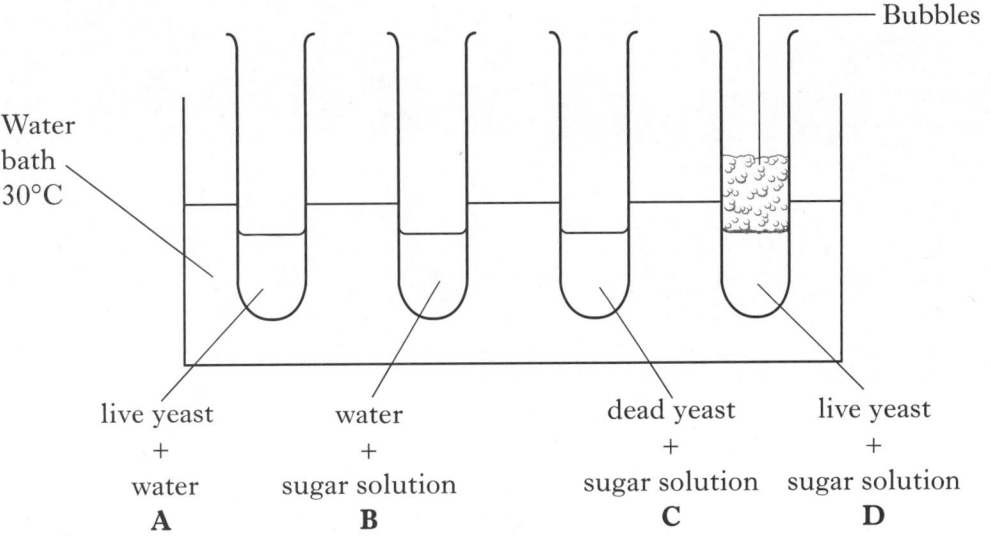

— Bubbles

Water
bath
30°C

live yeast   water   dead yeast   live yeast
+   +   +   +
water   sugar solution   sugar solution   sugar solution
A   **B**   **C**   **D**

(a) Explain why bubbles were formed only in test tube D.

_____

_____   1

(b) Predict what would happen to the volume of bubbles in test tube D if the experiment had been carried out in a water bath at 80 °C.

_____   1

(c) Which test tubes are controls?
Tick the correct box.

Tubes A and B only   ☐

Tubes B and C only   ☐

Tubes A and C only   ☐

Tubes A, B and C   ☐   1

[Turn over

Marks KU PS

18. The diagram below shows how a tomato grower produced tomato plants with a high yield of good fruits and a strong root system.

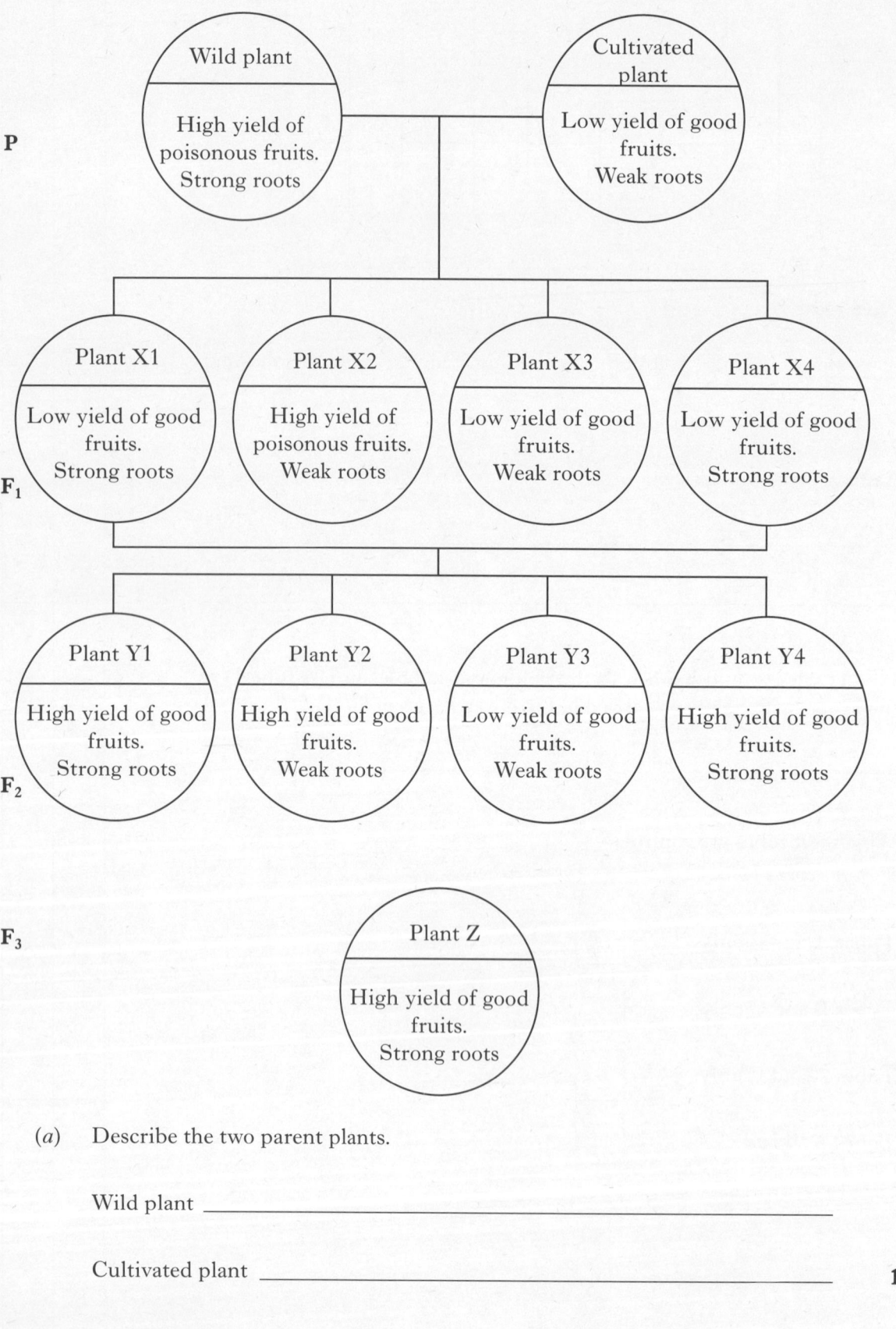

(a) Describe the two parent plants.

Wild plant _____

Cultivated plant _____   1

DO NOT WRITE IN THIS MARGIN

Marks | KU | PS

18. (continued)

(*b*) (i) Which two F_1 plants are shown being crossed to produce F_2 plants?

Plants __________ and __________

(ii) Explain why these plants were chosen.

____

____ **2**

(*c*) Which two F_2 plants should be used to obtain the generation of F_3 plants similar to Plant Z?

Plants __________ and __________ **1**

(*d*) What is the name given to this type of breeding programme?

______________________________ **1**

[Turn over

19. The graphs below show the results of tests on two enzymes for use in biological washing powders.

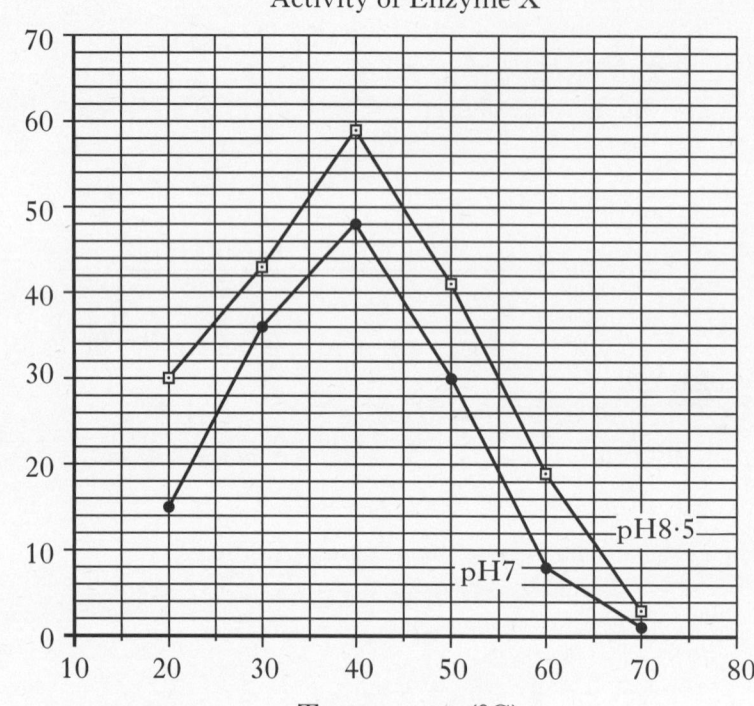

Activity of Enzyme X

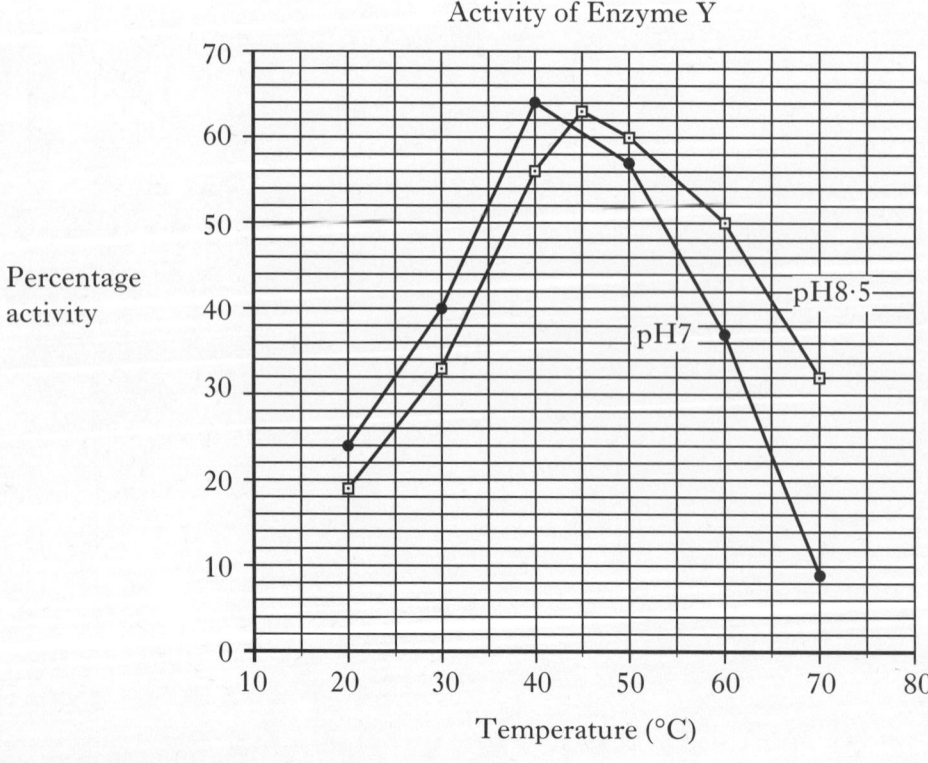

Activity of Enzyme Y

Marks KU PS

19. (continued)

(*a*) (i) Complete the table to show the activity of the two enzymes at pH 8·5 at different wash temperatures.

Type of wash	Enzyme X (% activity)	Enzyme Y (% activity)
Warm (40 °C)		56
Medium (50 °C)	41	60
Hot (60 °C)	19	

1

(ii) Most washing powders contain detergents that make the conditions alkaline, around pH 8 or 9. Which enzyme would be best to use for a hot wash?

Enzyme _____

1

(iii) Describe the effect of decreasing pH on the activity of Enzyme X.

_____

_____

1

(*b*) <u>Underline</u> **one** word in each group to make the sentences correct.

Enzymes are found in $\left\{ \begin{array}{l} some \\ most \\ all \end{array} \right\}$ cells and are made of $\left\{ \begin{array}{l} protein \\ carbohydrate \\ fat \end{array} \right\}$.

Enzymes are $\left\{ \begin{array}{l} substrates \\ reagents \\ catalysts \end{array} \right\}$ and work best in $\left\{ \begin{array}{l} hot \\ warm \\ cold \end{array} \right\}$ conditions.

2

[Turn over

Marks | KU | PS

20. The diagram below represents a sewage treatment plant.

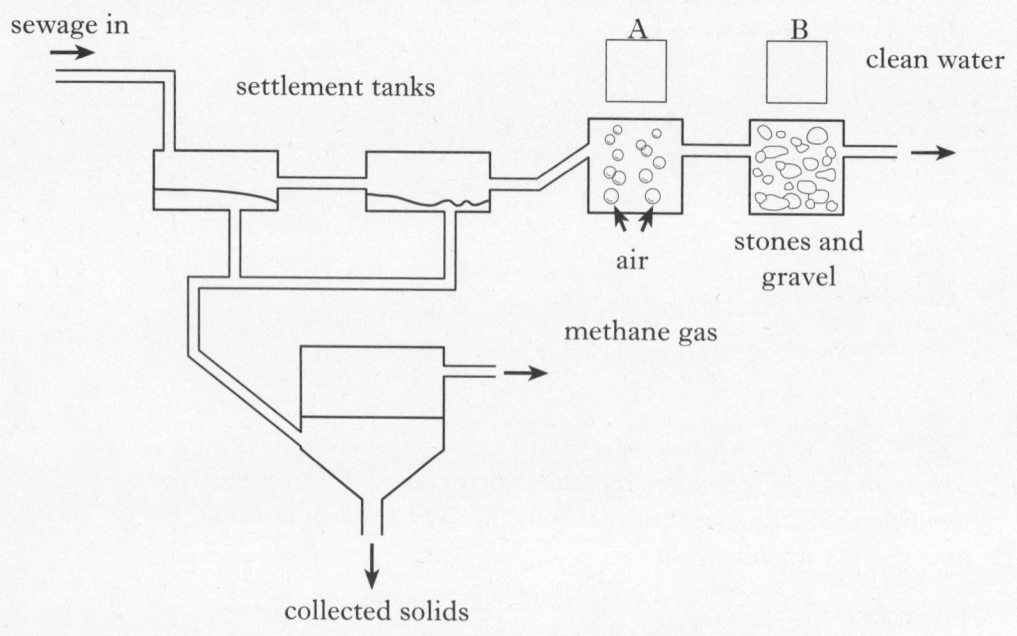

sewage in

settlement tanks

A   B   clean water

stones and
gravel

air

methane gas

collected solids

(*a*) What type of organisms are involved in the breakdown of sewage into harmless products during stages A and B?

_____

1

(*b*) The methane gas and the collected solids may be of economic importance.

Choose **one** of these products and explain its value.

Product _____

Value _____

_____

1

[END OF QUESTION PAPER]

SPACE FOR ANSWERS
AND FOR ROUGH WORKING

ADDITIONAL GRID FOR QUESTION 3(*a*)

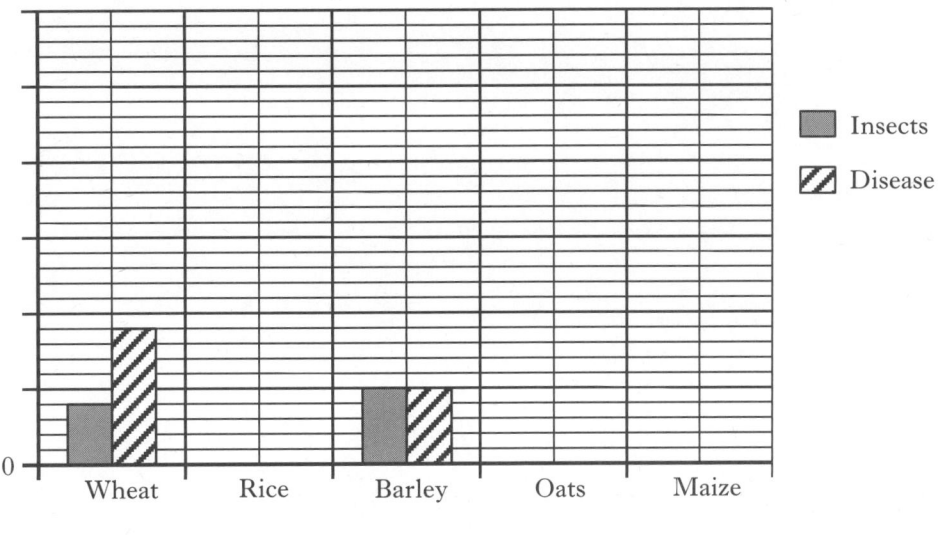

ADDITIONAL GRID FOR QUESTION 12(*a*)(i)

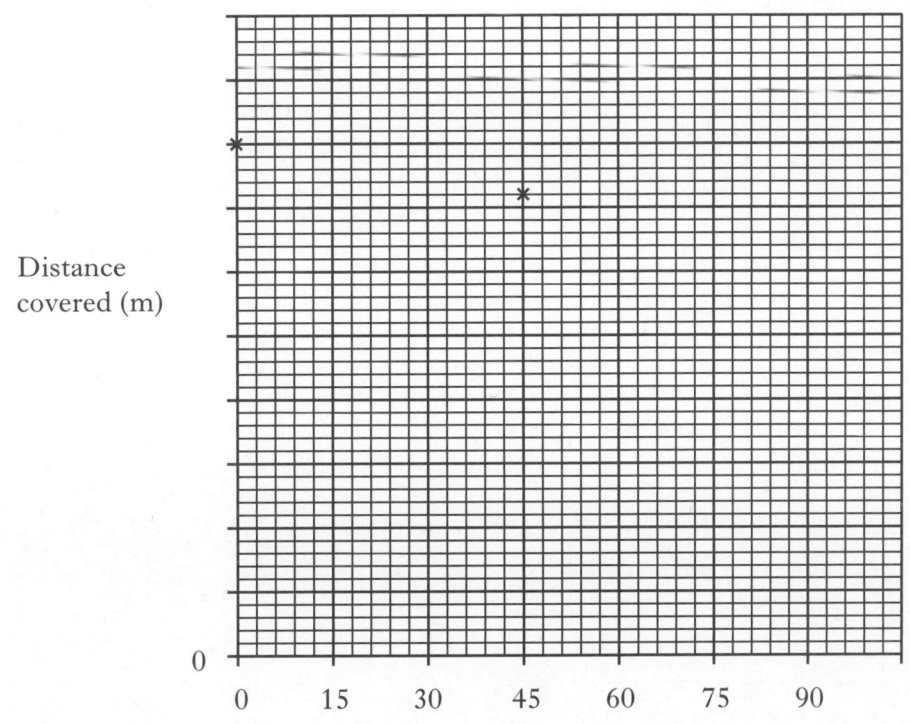

SPACE FOR ANSWERS
AND FOR ROUGH WORKING

[BLANK PAGE]

G

FOR OFFICIAL USE

KU   PS

Total Marks

0300/401

NATIONAL	FRIDAY 24 MAY	BIOLOGY
QUALIFICATIONS	9.00 AM – 10.30 AM	STANDARD GRADE
2002		General Level

Fill in these boxes and read what is printed below.

Full name of centre

Town

Forename(s)

Surname

Date of birth
Day Month Year   Scottish candidate number   Number of seat

1 All questions should be attempted.

2 The questions may be answered in any order but all answers are to be written in the spaces provided in this answer book, and must be written clearly and legibly in ink.

3 Rough work, if any should be necessary, as well as the fair copy, is to be written in this book. Additional spaces for answers and for rough work will be found at the end of the book. Rough work should be scored through when the fair copy has been written.

4 Before leaving the examination room you must give this book to the invigilator. If you do not, you may lose all the marks for this paper.

SCOTTISH
QUALIFICATIONS
AUTHORITY

Marks | KU | PS

1. A sports club wants to find out how well a weedkiller will get rid of dandelions on the rugby pitch. One area of the pitch was sampled using 1m² quadrats before spraying with the weedkiller and again three weeks after spraying. The results are shown below.

Quadrat	Number of dandelions	
	Before spraying	After spraying
1	3	1
2	5	1
3	1	0
4	4	2
5	7	2
6	2	1
7	6	2
8	3	2
9	5	2
10	4	2
Average number per m²		1·5

(a) (i) Complete the table by writing in the average number of dandelions per m² before spraying.

Space for calculation

1

(ii) The area sampled was 1000 m².

Calculate the estimated total number of dandelions present after spraying.

Space for calculation

Total _____

1

(b) How could the reliability of these results have been improved?

_____

1

DO NOT WRITE IN THIS MARGIN

Marks | KU | PS

1. (continued)

(*c*) (i) Name **two** abiotic factors that may affect the distribution of dandelions on the pitch.

1 ______________________________

2 ______________________________ **1**

(ii) Select **one** of the named abiotic factors and describe how you would measure it.

Factor ______________________________

Description ______________________________

______________________________ **1**

[Turn over

2. (*a*) The food of eight animals is listed in the table.

Animal	Food
beetles	oak bark
caterpillars	oak leaves
slugs	oak leaves
woodmice	oak bark
spiders	beetles, caterpillars
chaffinches	spiders, caterpillars, slugs
owls	woodmice
hawks	woodmice, chaffinches

(i) Use the information in the table to place each animal into the correct position on the food web below.

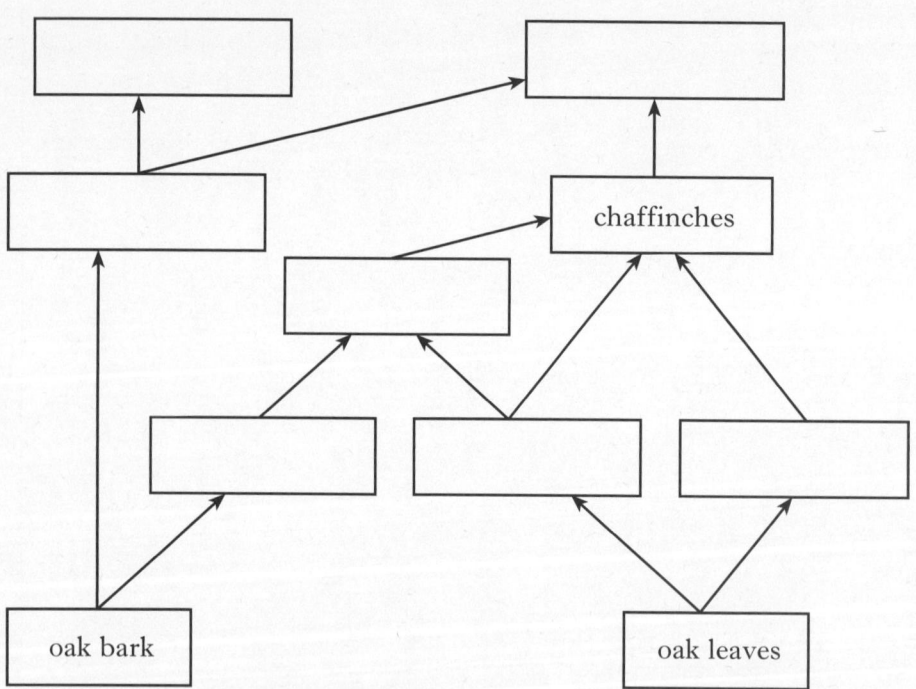

3

(ii) Choose **one** of the animals from the table and name or describe a suitable sampling technique.

Animal _____

Sampling technique _____

_____

1

Marks | KU | PS

2. (*a*) **(continued)**

(iii) The owls and the hawks are in competition with each other. Explain what this means.

_____   1

(iv) State **one** possible effect of competition between organisms.

_____

_____   1

(*b*) Complete the sentences below by using the correct words from the list.

List   community   producers   habitat

          population   biosphere   consumers

The place where an organism lives is its _____ .

All the members of one species living together are called a

_____ .

The _____ and habitats make up an ecosystem.   2

[**Turn over**

3. An experiment was set up to investigate the effect of light intensity on the rate of photosynthesis.

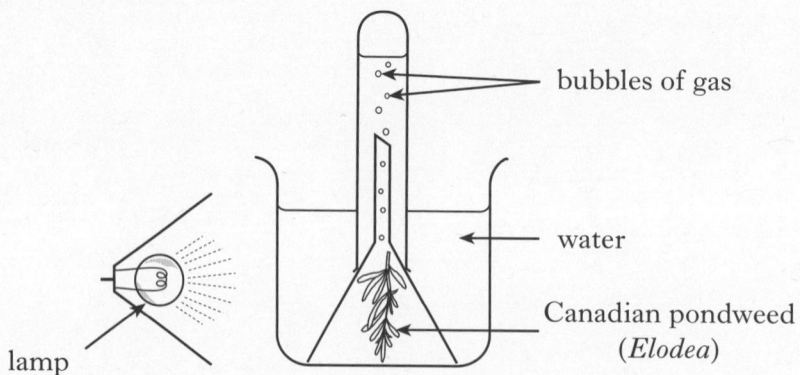

bubbles of gas

water

Canadian pondweed (*Elodea*)

lamp

The *Elodea* was exposed to different light intensities and the rate of photosynthesis was estimated by counting the number of bubbles of gas produced per minute. The results are shown below.

Light intensity (units)	0	1	2	3	4	5	6	7
Average number of bubbles per minute	0	7	14	20	25	27	27	27

(a) On the grid below, complete a **line graph** of the results by

(i) completing the vertical y-axis   1

(ii) putting a scale on the horizontal x-axis   1

(iii) plotting the graph.   1

(An additional grid, if needed, will be found on page 30.)

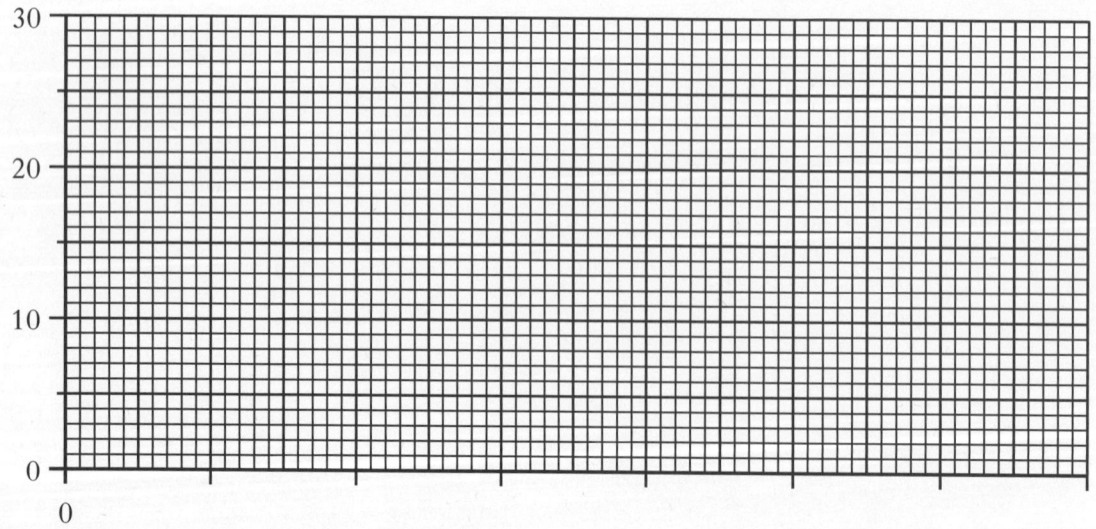

Light intensity (units)

DO NOT WRITE IN THIS MARGIN

Marks KU PS

3. (continued)

(*b*) Describe the effect on the rate of bubbling of increasing the light intensity **from 5 to 7 units**.

____ **1**

(*c*) Suggest a method for changing the light intensity in this experiment.

____ **1**

(*d*) The number of bubbles per minute at each light intensity was counted four times and an average calculated.

Explain why this was good experimental technique.

____ **1**

(*e*) Name the gas that forms the bubbles in this experiment.

________________ **1**

[Turn over

Official SQA Past Papers: General Biology 2002

DO NOT
WRITE IN
THIS
MARGIN

Marks | KU | PS

4. (*a*) Five tubes were set up as shown in the diagram below.

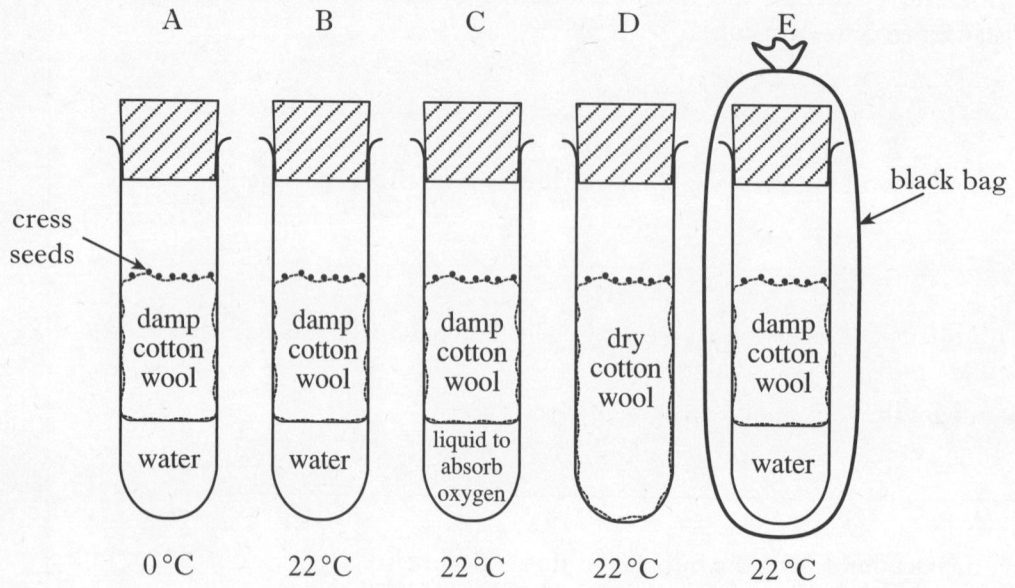

(i) In which **two** tubes would germination occur?
Tick the correct boxes.

A

B

C

D

E

(ii) Name the **four** factors being investigated in this experiment.

1 _____

2 _____

3 _____

4 _____

1

2

4. **(continued)**

(*b*) Stages in the reproduction of a flowering plant are named below.

flowering   fruit formation   fertilisation   pollination

Show the correct sequence of these stages by writing them in the appropriate boxes.

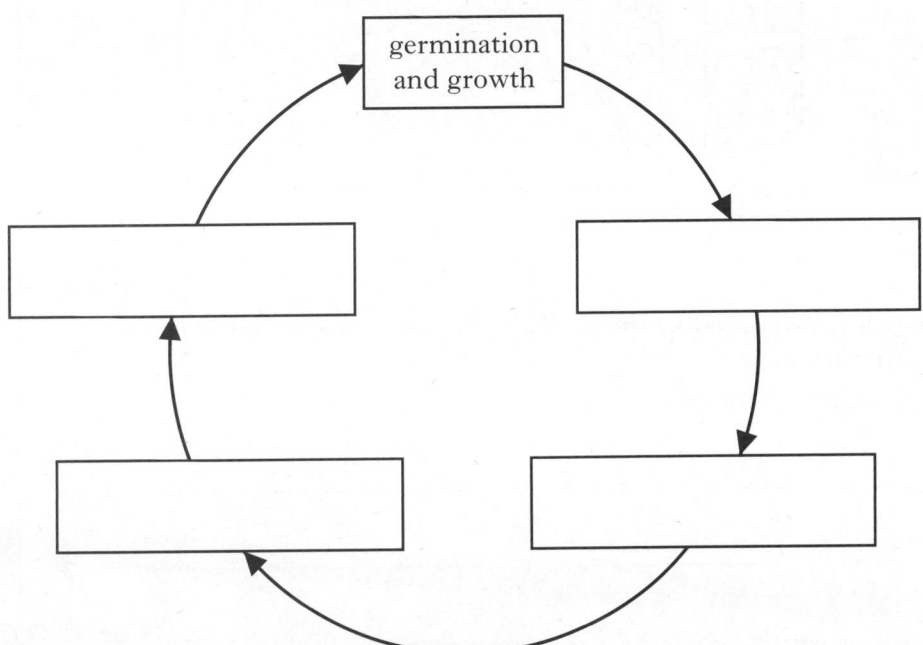

2

(*c*) Name **two** ways in which pollen can be transferred from one plant to another.

1 _____

2 _____

1

[Turn over

Marks | KU | PS

5. (*a*) The experiment below was set up to show that carbon dioxide is essential for photosynthesis.

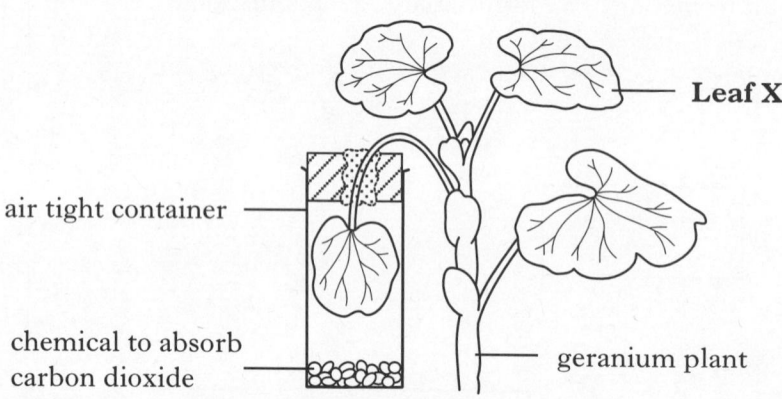

Leaf X

air tight container

chemical to absorb carbon dioxide

geranium plant

(i) The plant was placed in the dark for 24 hours before setting up the experiment.

Suggest a reason for this.

_____

_____ 1

(ii) Describe a suitable control for this experiment.

_____

_____ 1

(*b*) (i) Name the storage carbohydrate produced in **Leaf X** as a result of photosynthesis.

_____ 1

(ii) Complete the following sentence.

_____ is a green chemical, found in plant leaves,

that converts light energy into _____ energy

during photosynthesis. 1

Marks | KU | PS

6. The pie charts show the sources of Vitamins A and C in the diet.

Vitamin A

Vitamin C

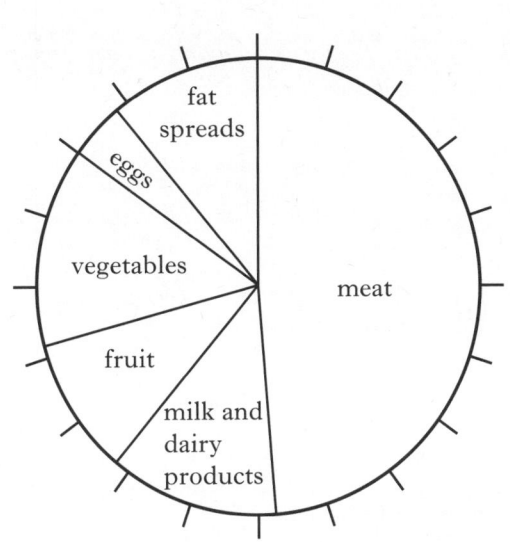

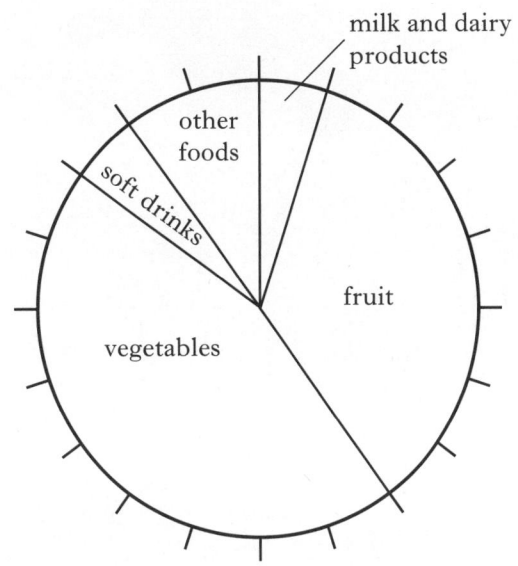

(a) Use the information from the pie charts to complete the table for Vitamin C.

Source of Vitamin A	Percentage of daily intake
Milk and dairy products	12
Fruit	10
Vegetables	14
Eggs	4
Fat spreads	11
Meat	49

Source of Vitamin C	Percentage of daily intake
Milk and dairy products	
Fruit	
Vegetables	
Soft drinks	
Other foods	

2

(b) What named foods supply the greatest proportion of

   (i) Vitamin A? _____

   (ii) Vitamin C? _____   1

(c) What **named** source of Vitamin C does not provide any Vitamin A?

_____   1

[Turn over

Official SQA Past Papers: General Biology 2002

DO NOT
WRITE IN
THIS
MARGIN

Marks | KU | PS

7. The diagram shows part of the digestive system of a rabbit.

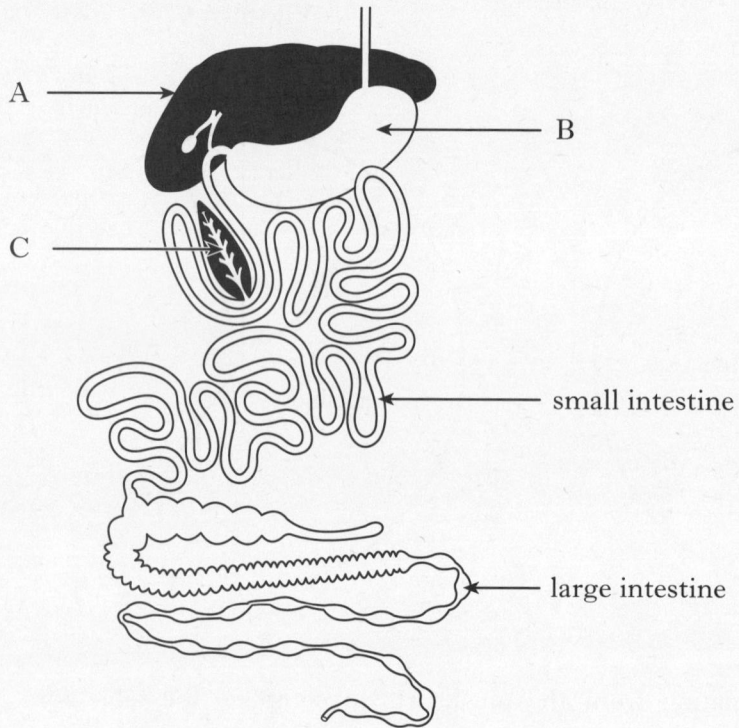

A

B

C

small intestine

large intestine

(*a*) Name the organs labelled on the diagram.

A _____

B _____

C _____   **2**

(*b*) In which part of the digestive system does most absorption of digested food occur?

_____   **1**

(*c*) Name the type of protein that carries out the reactions of digestion when mixed with food in the digestive system.

_____   **1**

(*d*) Food is required to provide animals with energy.
Name **one** other reason why food is required by animals.

_____   **1**

Marks | KU | PS

8. (a) Complete the table below by naming **one** organ that receives protection from each of the given parts of the skeleton.

Part of skeleton	Organ receiving protection
Skull	
Rib cage	
Backbone	

2

(b) Movement of the skeleton is caused by the contraction of muscles.
Name the structures that connect muscles to bones.

_____

1

(c) The following statements refer to experiments carried out on bone.

1. When a bone is soaked in acid for a few days it becomes soft and flexible.

2. When a bone is roasted it becomes hard and brittle.

Choose **one** of the statements and state which component of the bone has been removed by the experiment.

Statement number _____

Component removed _____

1

(d) The diagrams below represent a human arm.

Position A   Position B

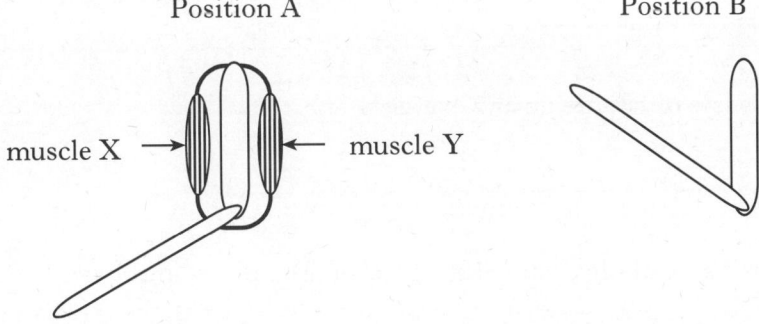

muscle X →   ← muscle Y

(i) Which muscle, X or Y, contracts to move the arm from position A to position B?

_____

1

(ii) Name the chemical that builds up in muscles which contract repeatedly for long periods.

_____

1

Marks KU PS

9. Read the passage below and answer the questions which follow it.

Designing a sports drink. Adapted from an article in *Biological Sciences Review*, September 2000.

Advertisements claim that athletes can improve their performance by drinking specially formulated sports drinks which reduce or delay fatigue. One of the causes of fatigue during exercise is the reduction of energy stores such as the glycogen in the muscles. Other causes of fatigue include problems associated with overheating and fluid loss.

During intense exercise the body mainly uses carbohydrate as an energy source. Taking carbohydrate during exercise can delay fatigue by conserving the energy stores in the muscles.

There is a dramatic increase in heat production during vigorous exercise. This does not result in a large increase in body temperature because heat is lost from the body by the evaporation of sweat from the skin surface. This reduces the risk of a raised body temperature but results in dehydration (a reduction in the body water content). Dehydration decreases performance during exercise by reducing the volume of blood available to meet the needs of all the tissues.

Most sports drinks have a similar composition—carbohydrate, water and sodium. Improvement in the taste and "mouthfeel" of drinks can be achieved by using different forms of carbohydrate such as glucose, fructose and sucrose. The sodium content of most sports drinks is normally less than half that in sweat largely for taste reasons.

(a) Name the energy store mentioned in the passage which is found in muscles.

_____ **1**

(b) State **one** of the causes of fatigue during exercise.

_____ **1**

(c) Explain why taking carbohydrate during exercise may improve performance.

_____ **1**

(d) Give **one** way in which sweating during exercise can

 (i) decrease fatigue _____ **1**

 (ii) increase fatigue. _____ **1**

Marks KU PS

9. **(continued)**

(e) State the **three** main components of sports drinks.

1 _____

2 _____

3 _____   1

(f) Why do sports drinks contain less than half the sodium content of sweat?

_____   1

[Turn over

Page fifteen

Marks | KU | PS

10. The diagram shows the life cycle of the Atlantic Salmon. The salmon are able to migrate between their breeding grounds in Scottish rivers and their feeding grounds in the Atlantic Ocean. Adult salmon migrate between the rivers and the ocean every year.

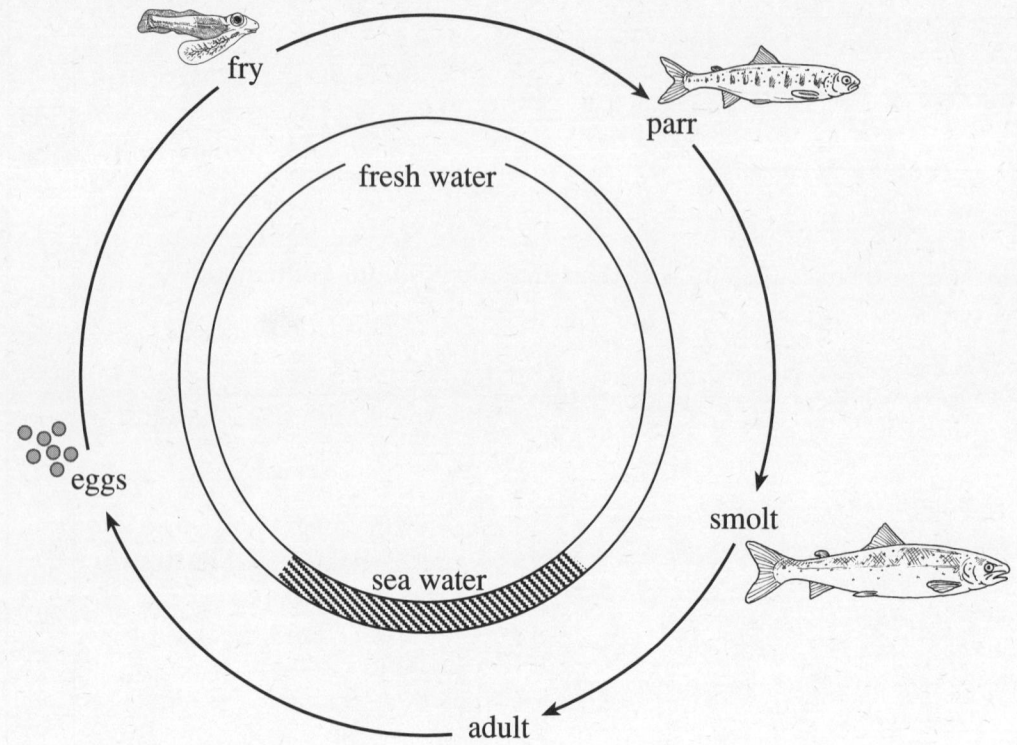

(a) What term is used to describe regular repeated behaviour patterns, such as the migration of the salmon?

_____

1

(b) Sea lice are a pest of adult salmon. Suggest why they never attack fry or parr.

_____

1

(c) From where do the young fry obtain their food?

_____

1

10. **(continued)**

(d) A female salmon lays 8000 eggs but only 5% of them hatch.
How many fry will be produced?
Space for calculation

Number of fry _____

1

(e) The table shows the number of salmon caught in a Scottish river over a six year period.

Month	Number of fish caught					
	1991	1992	1993	1994	1995	1996
May	1	9	15	3	0	0
June	103	125	139	109	171	234
July	207	390	267	225	216	276
August	76	168	159	103	72	48
September	17	57	41	13	21	1
Total	404	749	621	453	480	559

(i) During which month were the greatest number of fish caught?

_____

1

(ii) What percentage was the August catch of the total for 1995?
Space for calculation

_____ %

1

(iii) What was the average number of fish caught during September over the six year period?
Space for calculation

_____

1

Marks | KU | PS

11. (*a*) <u>Underline</u> **one** word in each set of brackets to make the sentences correct.

$\left\{\begin{array}{l} Cells \\ Tissues \\ Organs \end{array}\right\}$ are the basic units of living organisms.

Most of them are too small to be seen with the naked eye and are almost transparent. A microscope magnifies them so that we can see them, and coloured chemicals called

$\left\{\begin{array}{l} indicators \\ stains \\ pigments \end{array}\right\}$ can be added to make certain parts easier to see.

1

(*b*) The following are drawings of cells. They are not drawn to the same scale.

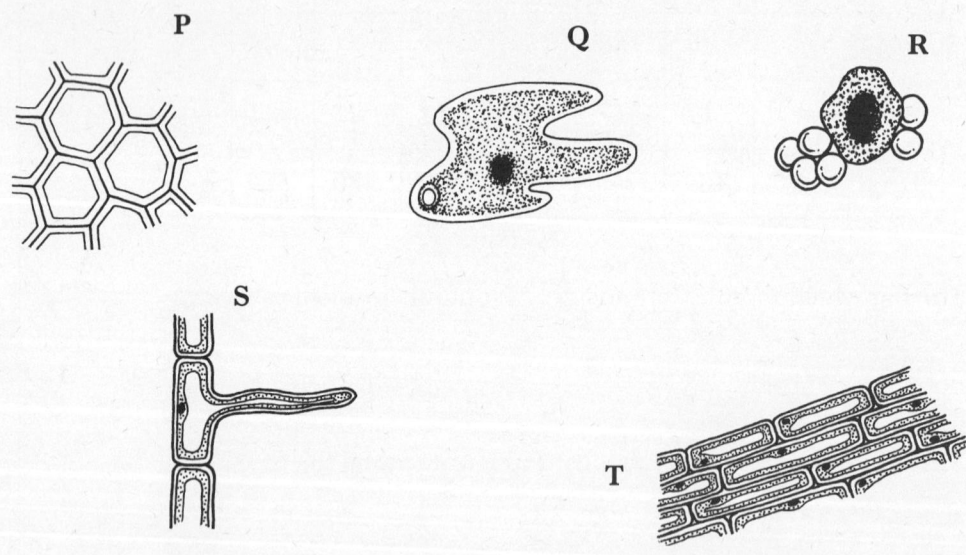

P   Q   R

S

T

Give the letters of **all** the animal cells.

Animal cells _____

1

11. **(continued)**

(c) (i) Complete the following sentence to give a definition of the process of diffusion.

Diffusion is the movement of a substance

from an area of _____

to an area of _____ .   **1**

(ii) The list below names three substances which diffuse into and out of living cells.

List   dissolved food   carbon dioxide   oxygen

Complete the diagram to show correctly the movement of each named substance into or out of the cell.

2

(iii) Which part of the cell controls the passage of substances into or out of the cell?

_____   **1**

(iv) What name is given to the "special case" of the diffusion of water into or out of cells?

_____   **1**

Official SQA Past Papers: General Biology 2002

DO NOT
WRITE IN
THIS
MARGIN

Marks | KU | PS

12. The graph shows the time from fertilisation to hatching of trout eggs kept at different temperatures at a Scottish trout farm.

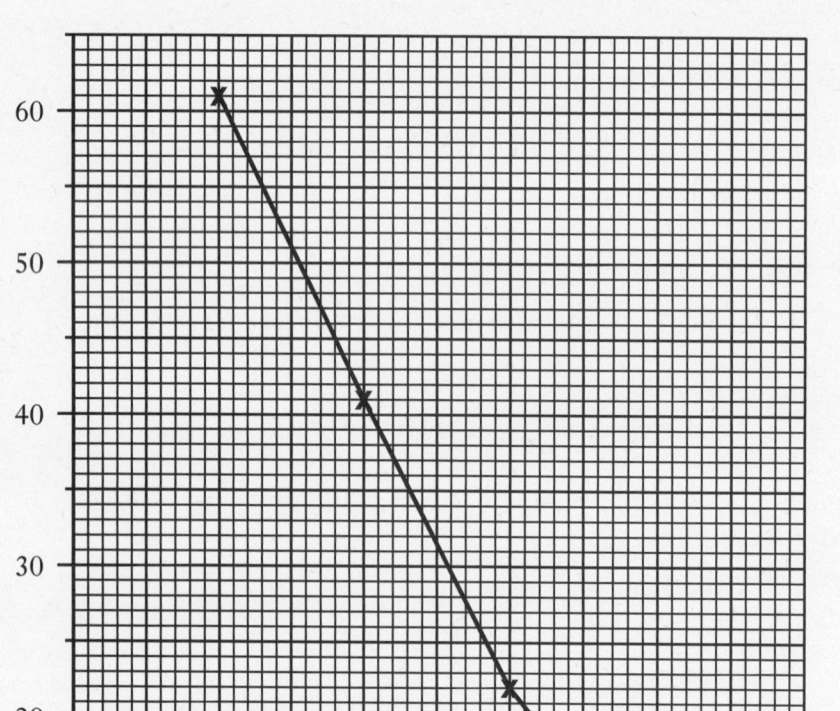

(a) (i) Give the length of time for the eggs to hatch at 10 °C.

_____ days

1

(ii) Describe the effect of increasing the temperature on the time to hatching.

_____

_____

1

DO NOT WRITE IN THIS MARGIN

Marks | KU | PS

12. (*a*) (continued)

(iii) Predict the effect that raising the temperature of the water to 50 °C would have on the hatching of the eggs. Give a reason for your answer.

Effect ______________________________ 1

Reason ______________________________

______________________________ 1

(*b*) (i) The trout eggs would not hatch if it were not for the presence of enzymes to act as catalysts.

Explain the meaning of the term "catalyst".

____

______________________________ 1

(ii) Give the name of **one** enzyme involved in the chemical breakdown of a substance and **one** enzyme involved in synthesis (build up).

Breakdown ______________________________ 1

Synthesis ______________________________ 1

[Turn over

13. The diagram represents stages in sexual reproduction of mammals.

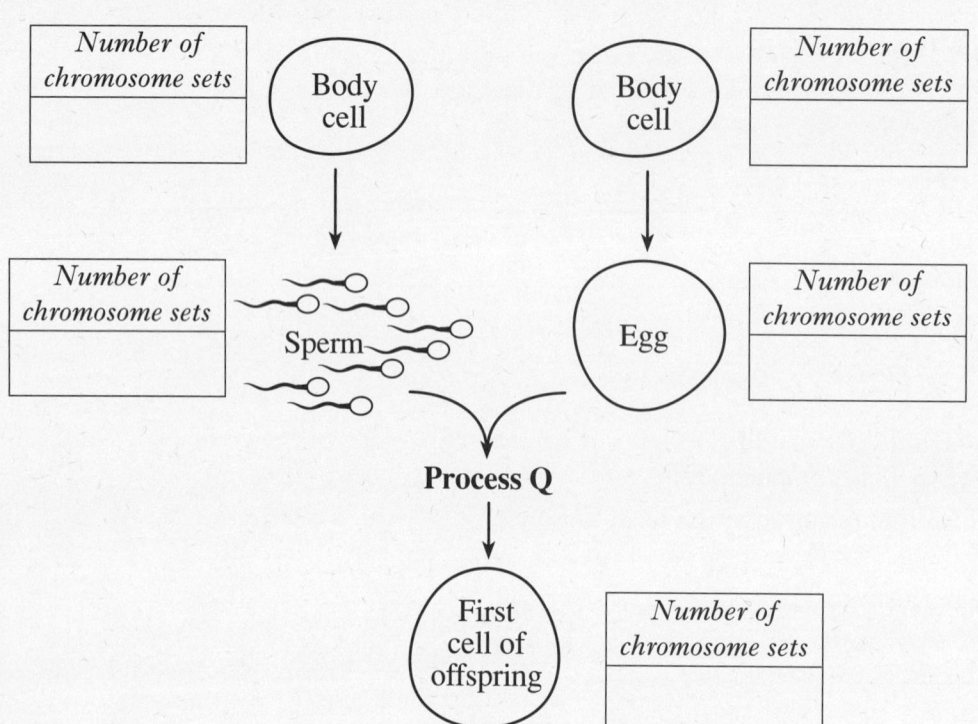

(a) Complete the diagram by writing in each box the number of **complete chromosome sets** for each cell type shown.

2

(b) What general name is given to sex cells such as eggs and sperm?

_____

1

(c) What name is given to **Process Q** in the diagram?

_____

1

13. **(continued)**

(*d*) The sex of an offspring is determined by the sex chromosomes **X** and **Y**. Complete each diagram below to identify the sex chromosomes present and the sex of the offspring **2** and **3**.

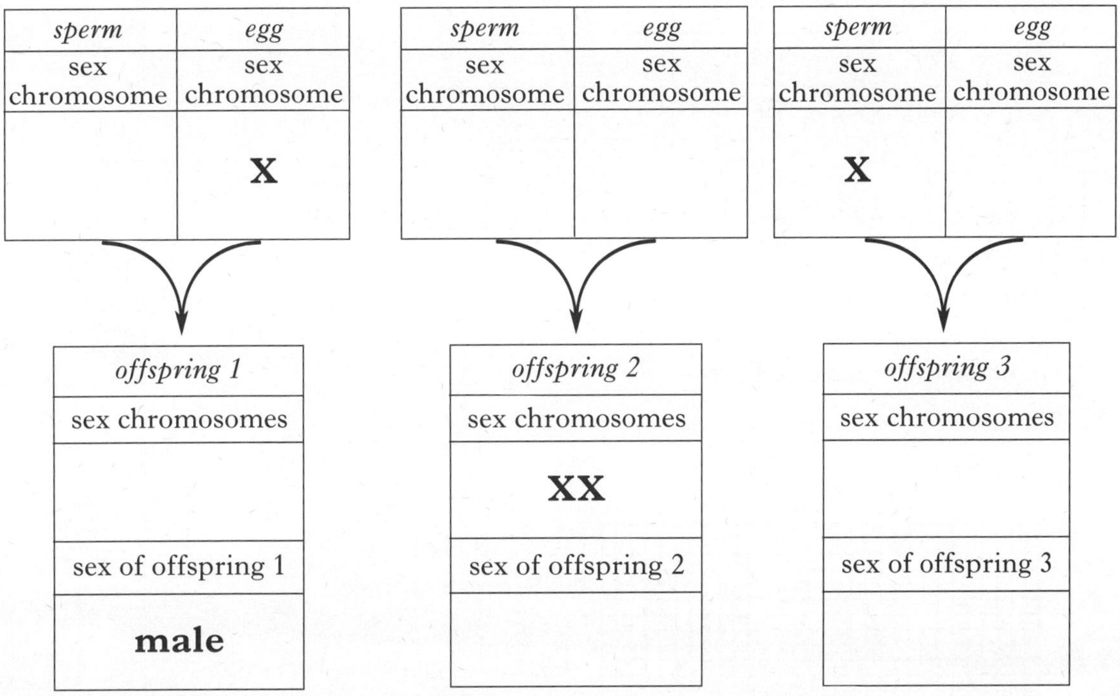

sperm	*egg*
sex chromosome	sex chromosome
	X

sperm	*egg*
sex chromosome	sex chromosome

sperm	*egg*
sex chromosome	sex chromosome
X	

offspring 1
sex chromosomes
sex of offspring 1
male

offspring 2
sex chromosomes
XX
sex of offspring 2

offspring 3
sex chromosomes
sex of offspring 3

3

[Turn over

Marks | KU | PS

14. (a) The table below shows the percentage of household waste recycled in Strathclyde compared to the overall Scottish average over a three year period.

Year	Household waste recycled (%)	
	Strathclyde	Scottish average
1990	1·4	2·2
1991	1·5	3·0
1992	1·6	3·2

(i) Use the information from the table to complete the bar chart below by

1 adding a scale to the y-axis

1

2 plotting the remaining bars.

1

(An additional grid, if needed, will be found on page 30.)

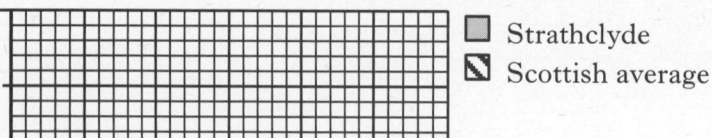

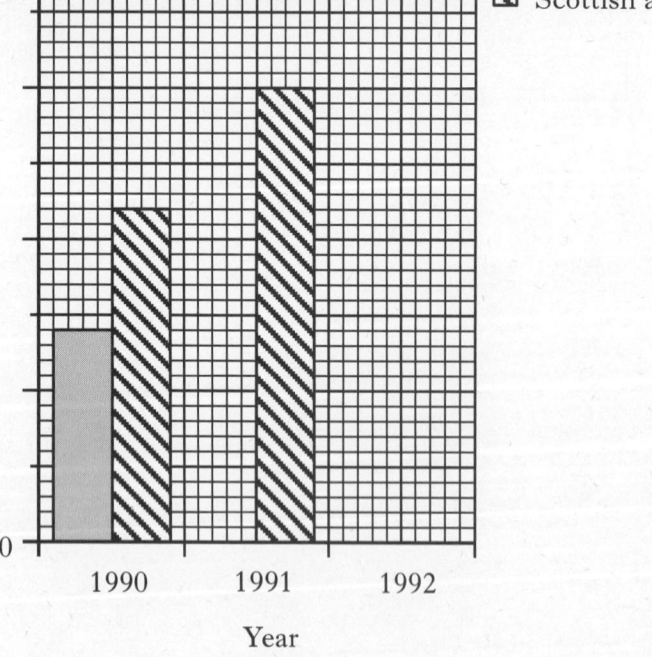

Household waste recycled (%)

Year

(ii) Express as a single whole number ratio the proportion of household waste recycled in Strathclyde to the Scottish average, in 1992.

Space for calculation

1

_____ : _____
Strathclyde   Scottish average

DO NOT WRITE IN THIS MARGIN

Marks	KU	PS

14. (*a*) (continued)

(iii) Describe the trend in the recycling of household waste in Scotland.

____ **1**

(*b*) Household waste is a domestic pollutant which can damage land ecosystems.

Sewage is another example of a domestic pollutant.

(i) Name the ecosystem which may be damaged by the discharge of untreated sewage.

______________________ **1**

(ii) Name a disease that can be spread by untreated sewage.

______________________ **1**

[Turn over

Marks | KU | PS

15. (a) The graph shows the number of bacteria in a liquid nutrient over 24 hours.

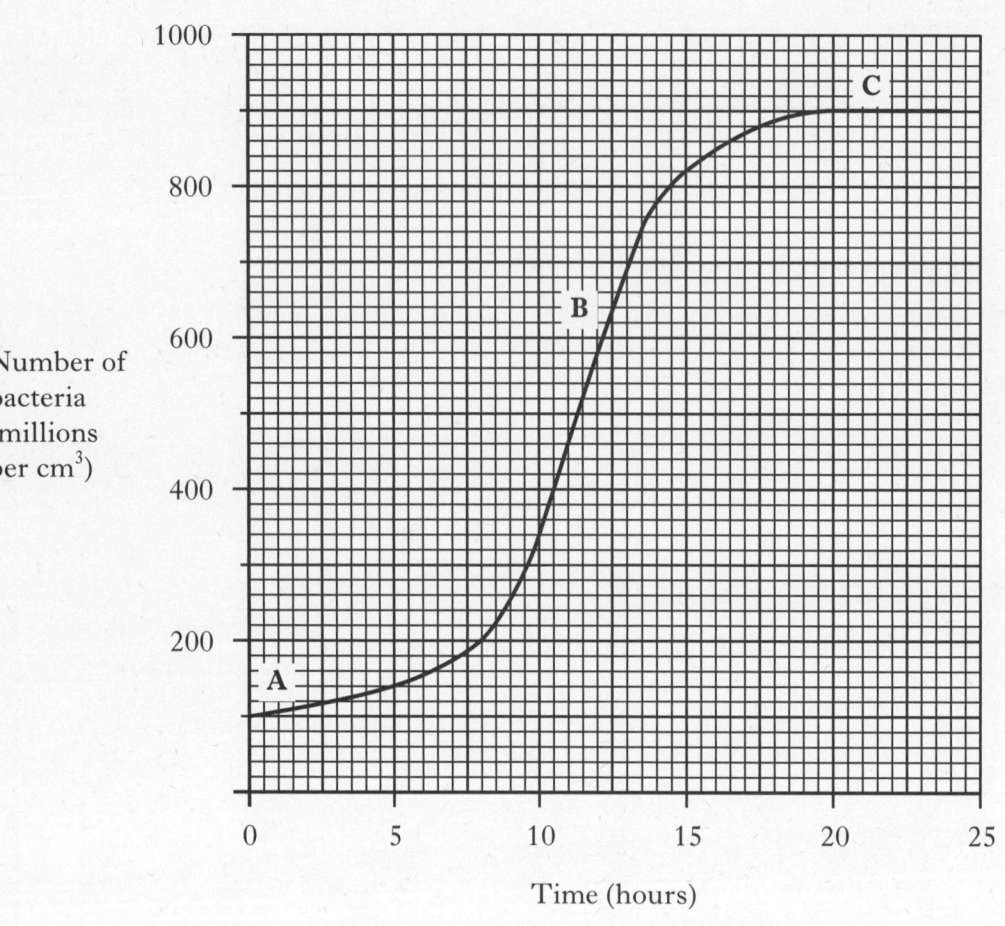

(i) How many bacteria were present at eight hours?

_____ millions per cm^3

1

(ii) Identify the stages at which the number of new bacteria being produced is greater than the number dying.

Stages _____ and _____

1

(iii) State **two** factors that could limit the growth of the population at stage C on the graph.

1 _____

2 _____

2

Marks | KU | PS

16. A group of pupils investigated the activity of yeast by making some dough with flour, water, sugar and yeast.

Enough water was added to make the dough runny and 50 cm^3 was poured into a beaker.

The volume of the dough was noted every 10 minutes and the results are shown in the table.

Time *(minutes)*	0	10	20	30	40
Volume of dough (cm^3)	50	54	62	74	80

(a) (i) What was the increase in the volume of the dough during the time of this investigation?

_____ cm^3

1

(ii) Express this increase as a percentage of the original volume.

Space for calculation

_____ %

1

(b) During which period was there the greatest increase in the volume of the dough?

Tick the correct box

☐ 0–10 minutes

☐ 10–20 minutes

☐ 20–30 minutes

☐ 30–40 minutes

1

(c) The teacher suggested that the results might not be typical but could be unusual or a "one-off" result. How could the investigation be improved to overcome this problem?

_____

1

DO NOT WRITE IN THIS MARGIN

KU PS

16. (continued)

(*d*) One of the group suggested that the raising of the dough might not be caused by the yeast but by some other factor.

(i) Describe another experiment which could be set up to test this idea.

____

____ **1**

(ii) What name is given to an experiment which is designed to make sure that the result of an investigation is only due to the factor being investigated?

______________________ **1**

[*END OF QUESTION PAPER*]

SPACE FOR ANSWERS
AND FOR ROUGH WORKING

ADDITIONAL GRID FOR QUESTION 3(*a*)

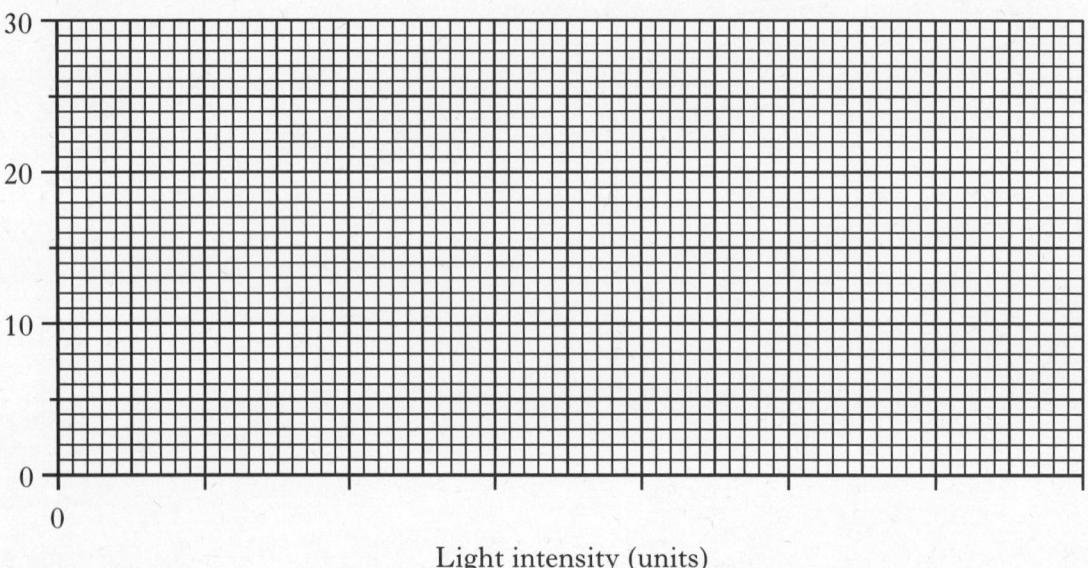

Light intensity (units)

ADDITIONAL GRID FOR QUESTION 14(*a*)(i)

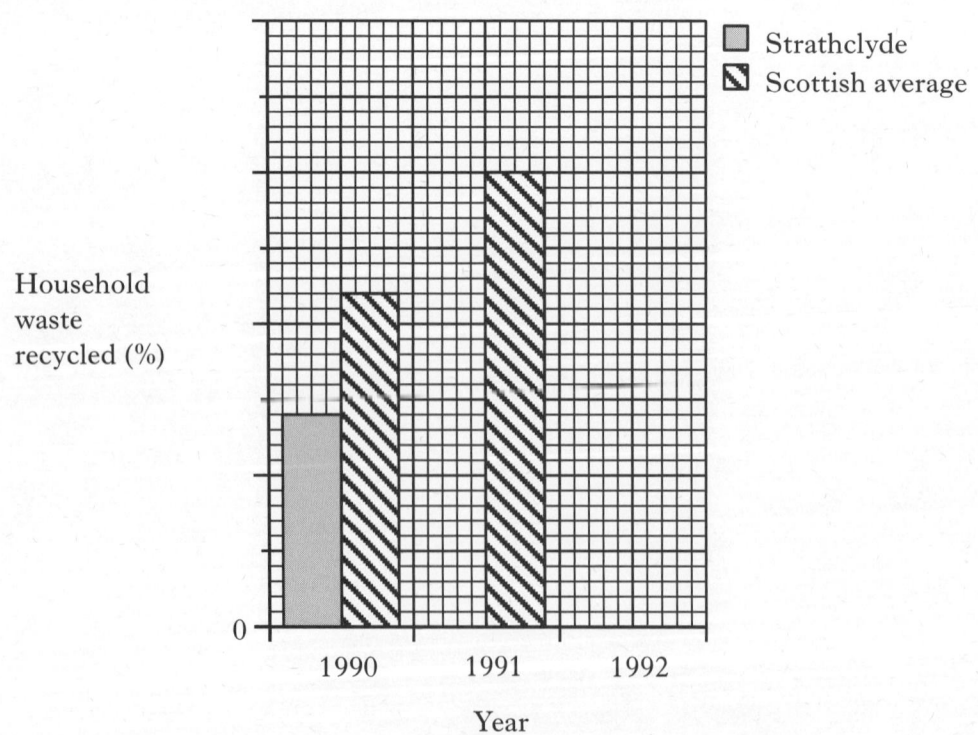

SPACE FOR ANSWERS
AND FOR ROUGH WORKING

ADDITIONAL PIE CHART FOR QUESTION 15(b)(i)

Percentage composition of single-cell protein

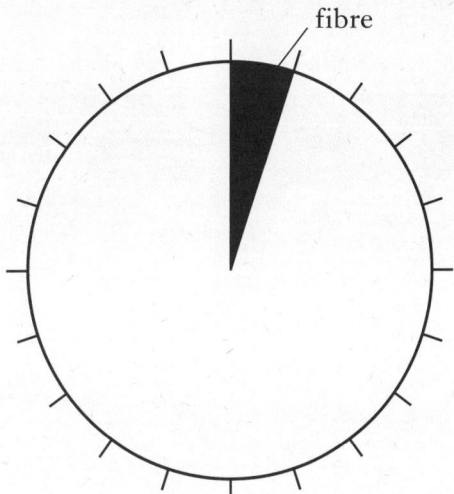

SPACE FOR ANSWERS
AND FOR ROUGH WORKING

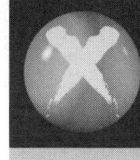

[BLANK]

FOR OFFICIAL USE

G

KU   PS

Total Marks

0300/401

NATIONAL
QUALIFICATIONS
2003

MONDAY 26 MAY
9.00 AM – 10.30 AM

**BIOLOGY
STANDARD GRADE**
General Level

Fill in these boxes and read what is printed below.

Full name of centre

Town

Forename(s)

Surname

Date of birth
Day   Month   Year   Scottish candidate number   Number of seat

1 All questions should be attempted.

2 The questions may be answered in any order but all answers are to be written in the spaces provided in this answer book, and must be written clearly and legibly in ink.

3 Rough work, if any should be necessary, as well as the fair copy, is to be written in this book. Additional spaces for answers and for rough work will be found at the end of the book. Rough work should be scored through when the fair copy has been written.

4 Before leaving the examination room you must give this book to the invigilator. If you do not, you may lose all the marks for this paper.

SCOTTISH
QUALIFICATIONS
AUTHORITY

DO NOT WRITE IN THIS MARGIN

Marks | KU | PS

1. The diagram gives some information about a woodland in southern Scotland.

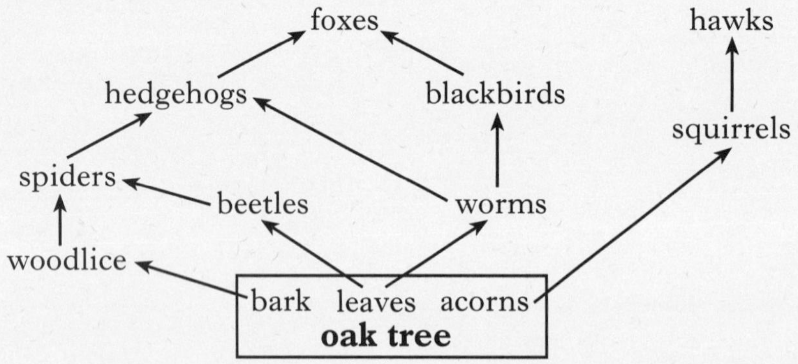

foxes

hawks

hedgehogs   blackbirds

squirrels

spiders ← beetles   worms

woodlice ← bark leaves acorns
oak tree

(a) What name is given to this type of diagram?

_____   1

(b) Answer the following using information **from the diagram**.

(i) Name **one** producer and **one** consumer.

Producer _____   Consumer _____   1

(ii) What do the arrows in the diagram represent?

_____   1

(iii) Complete the food chain below.

oak leaves ⟶ _____ ⟶ _____ ⟶ _____ ⟶ foxes   1

(iv) Name the part of the oak tree not involved in the food chains that include foxes.

_____   1

(v) Which part of the oak tree provides energy for the greatest number of different species?

_____   1

(c) Complete the table of words about the biosphere and their meanings.

Word	Meaning
habitat	
	all the animals or plants of a single species living in an area
	a particular area and all the animals and plants which live there

3

Marks | KU | PS

2. The table shows the mass of some of the main air pollutants produced in Britain in one year.

Pollutant	Mass produced (tonnes per year)
sulphur dioxide	4000
dust and grit	1500
carbon monoxide	6000
smoke	1000
others	500
TOTAL	

(a) Complete the table by entering the total mass of pollutants in the space provided.

1

(b) The pie chart below shows the information from the table.

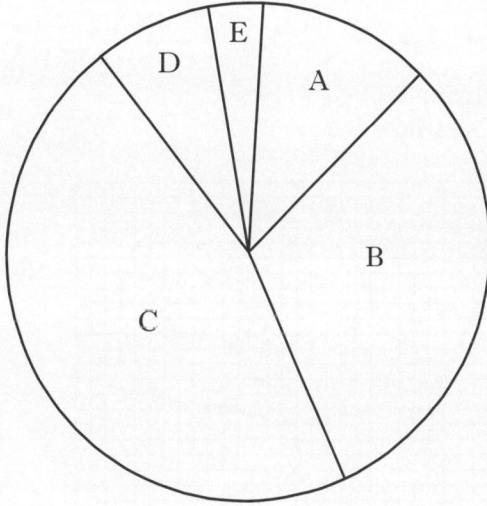

(i) Which letter represents the pollution due to dust and grit?

Letter _____

1

(ii) Identify the pollutants represented by segments C and D on the chart.

C _____

1

D _____

1

Marks | KU | PS

3. The graphs below show the oxygen concentrations upstream and downstream from the outflow pipes of two different sewage works, **A** and **B**. The two sewage works receive equal quantities of sewage and the two rivers are of equal size and speed.

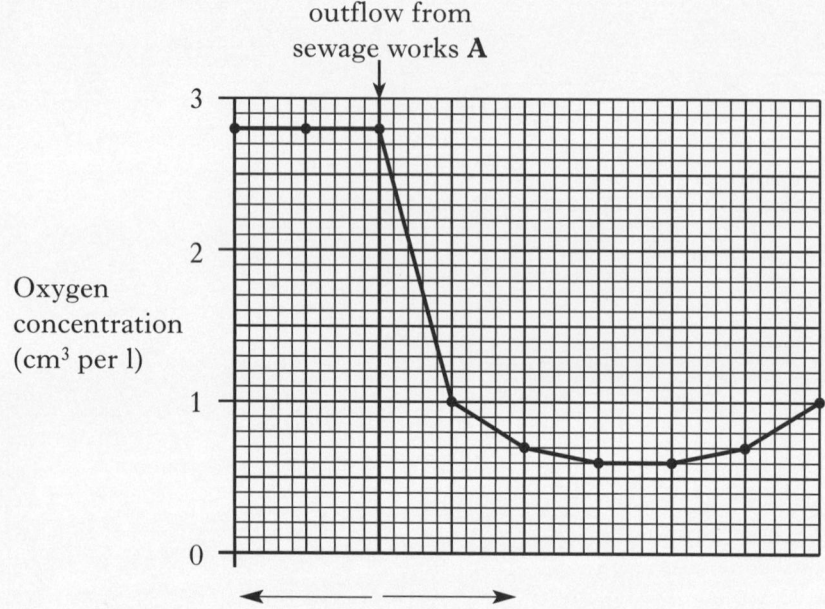

outflow from
sewage works **A**

Oxygen
concentration
(cm³ per l)

distance upstream   distance downstream

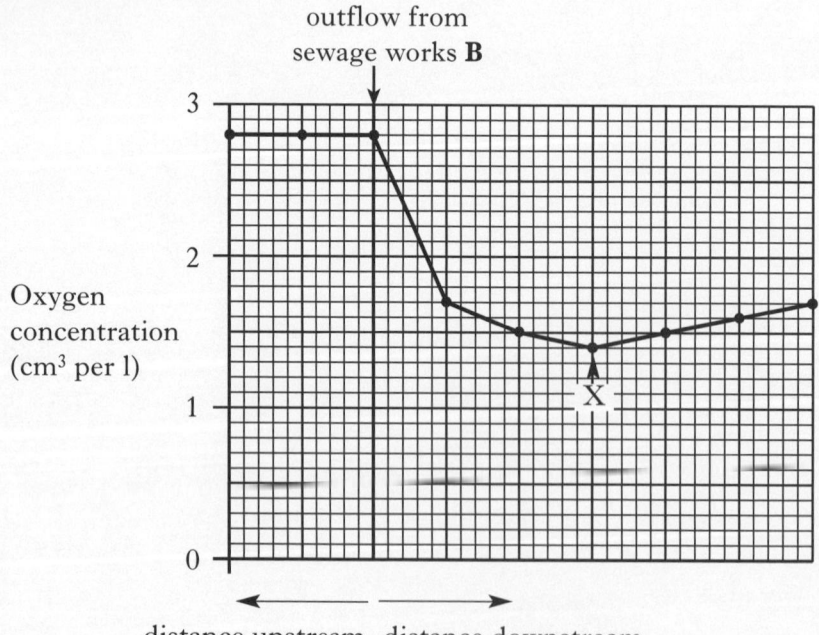

outflow from
sewage works **B**

Oxygen
concentration
(cm³ per l)

distance upstream   distance downstream

(*a*) What is the oxygen concentration of the water upstream from sewage works **A**?

_____ cm³ per l

1

3. **(continued)**

(b) Calculate the percentage of oxygen lost from the water between the outflow of sewage works **B** and point **X**.

Space for calculation

_____ %

1

(c) Complete the following sentence to describe the change in oxygen concentration which takes place downstream from both sewage works.

As the distance downstream from the sewage works increases, the

oxygen concentration _____ and then _____ .

1

(d) Which sewage works is more efficient at removing waste material from the sewage?

Give a reason for your answer.

Sewage works _____

1

Reason _____

_____

1

(e) Give **one** example of a disease that may be spread by untreated sewage.

_____

1

[Turn over

Marks | KU | PS

4. The diagrams show three different types of human teeth.

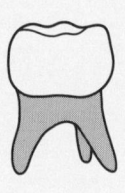      

A   B   C

(*a*)   (i)   Complete the following table by choosing **one** of the teeth, **A**, **B** or **C**, for each description.

Each letter may be used once, more than once or not at all.

Description of tooth	Tooth
Found at the very back of the jaw	
Known as an incisor	
Used for grinding and crushing food	

2

(ii)   State **one** function of canine teeth in carnivores.

_____

_____

1

Marks | KU | PS

4. (continued)

(b) Fluoride can be added to water supplies to help reduce tooth decay. The following bar graph shows the results of a study into the effect of some fluoride concentrations on the decay of children's teeth.

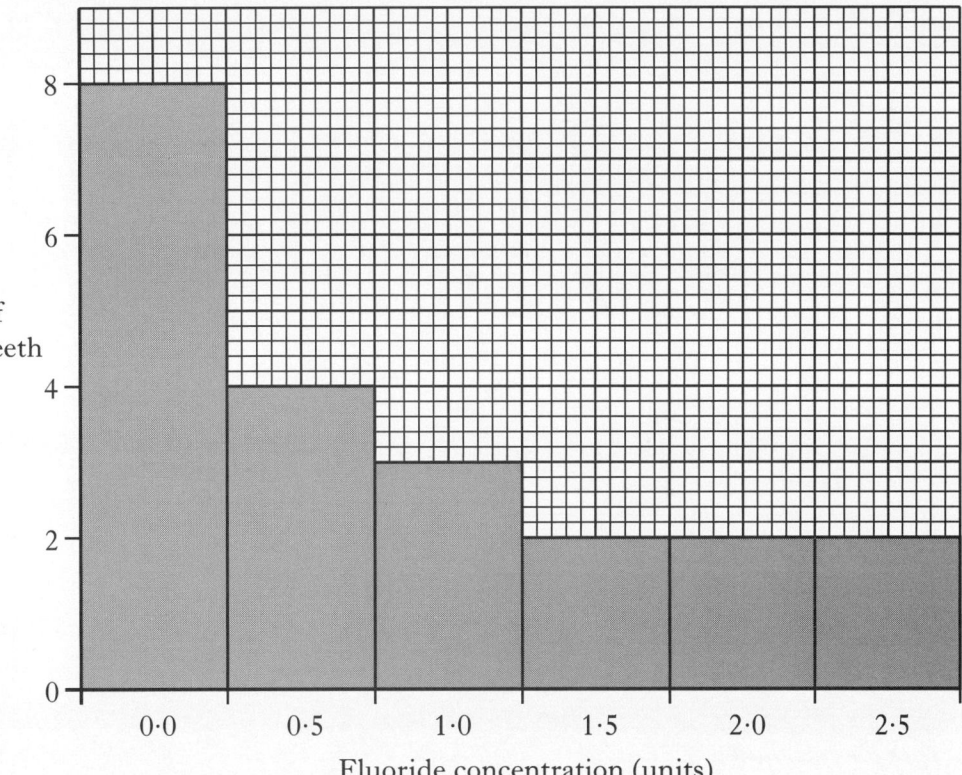

(i) What was the average number of decayed teeth per child when the fluoride concentration was 1 unit?

_____ decayed teeth per child

1

(ii) On average, how many teeth per child were saved from decay by increasing the fluoride concentration from 0·0 to 0·5 units?

_____ teeth per child

1

(iii) It may be concluded from the study that a fluoride concentration of 1·5 ppm is best.

Explain why this concentration would be better than each of the following.

1 1·0 unit _____

_____

1

2 2·0 units _____

_____

1

Marks KU PS

5. (*a*) The following diagram shows the main methods of water gain and loss for the human body.

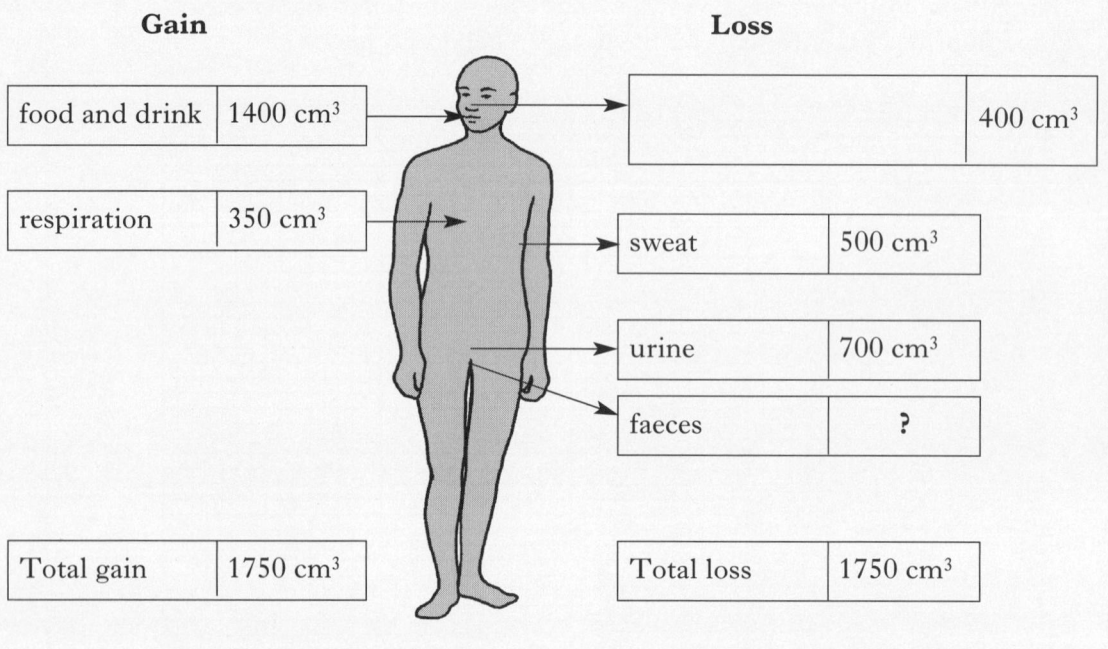

Gain

| food and drink | 1400 cm³ |
| respiration | 350 cm³ |

Loss

	400 cm³
sweat	500 cm³
urine	700 cm³
faeces	?

| Total gain | 1750 cm³ |

| Total loss | 1750 cm³ |

(i) Complete the empty box to name the missing method of water loss.

1

(ii) Calculate the volume of water lost in faeces.
Space for calculation

_____ cm³

1

(iii) What percentage of the water gained comes from respiration?
Space for calculation

_____ %

1

(*b*) Which organs are directly responsible for regulating the water content of the blood?

_____

1

(*c*) Name the poisonous waste substance that is removed in the urine together with water and salts.

_____

1

Marks | KU | PS

5. **(continued)**

(d) The table shows the volumes of juices released into the digestive system each day.

Digestive juice	Volume (cm³)
saliva	1500
gastric juice	2500
bile	500
pancreatic juice	700
intestinal juice	3000

Use the table to complete the **bar chart** below by:

(i) labelling the vertical axis   1

(ii) adding the scale to the vertical axis   1

(iii) completing the bars   1

(Additional graph paper, if required, will be found on page 27.)

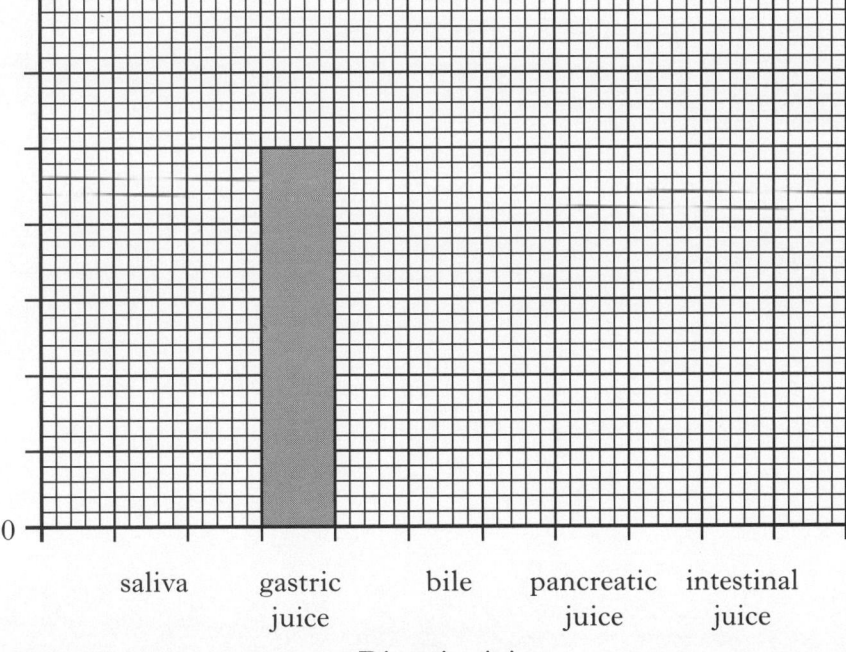

saliva   gastric juice   bile   pancreatic juice   intestinal juice

Digestive juice

(e) Which part of the digestive system reabsorbs most of the water from the juices?

_____   1

15. **(continued)**

(b) The composition of one type of single-cell protein is shown below.

Component	Percentage
protein	45
fat	10
minerals	5
fibre	5
other nutrients	35

(i) Use the information from the table to complete the pie chart below.

(An additional pie chart, if needed, will be found on page 31.)

Percentage composition of single-cell protein

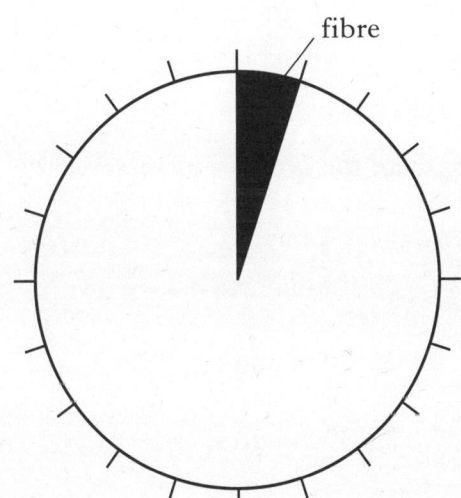

fibre

2

(ii) Calculate the ratio of protein to fat in the single-cell protein.
Express your answer as a simple whole number ratio.

Space for calculation

_____ : _____

protein   fat

1

(c) Micro-organisms are grown for the production of single-cell protein.
Name the type of reproduction involved.

_____

1

Marks | KU | PS

6. The diagram shows the lower surface of a leaf.

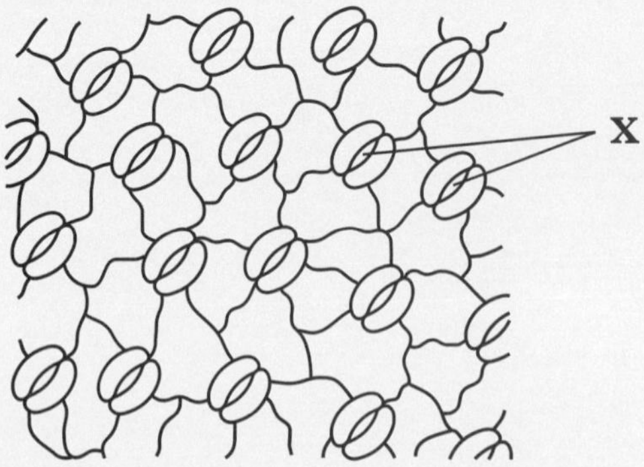

(a) (i) Name the pores labelled **X** on the diagram.

_____

1

(ii) Which gas, needed for photosynthesis, is taken in through these pores?

_____

1

(iii) The pores are able to open and close. Which substance, important for the growth of the plant, is conserved when the pores are closed?

_____

1

(b) During photosynthesis green plants produce glucose. This can be changed to an insoluble carbohydrate for storage. What is the name of this storage carbohydrate?

_____

1

(c) Name the green substance needed for photosynthesis.

_____

1

7. The diagrams show some asexual methods of plant reproduction.

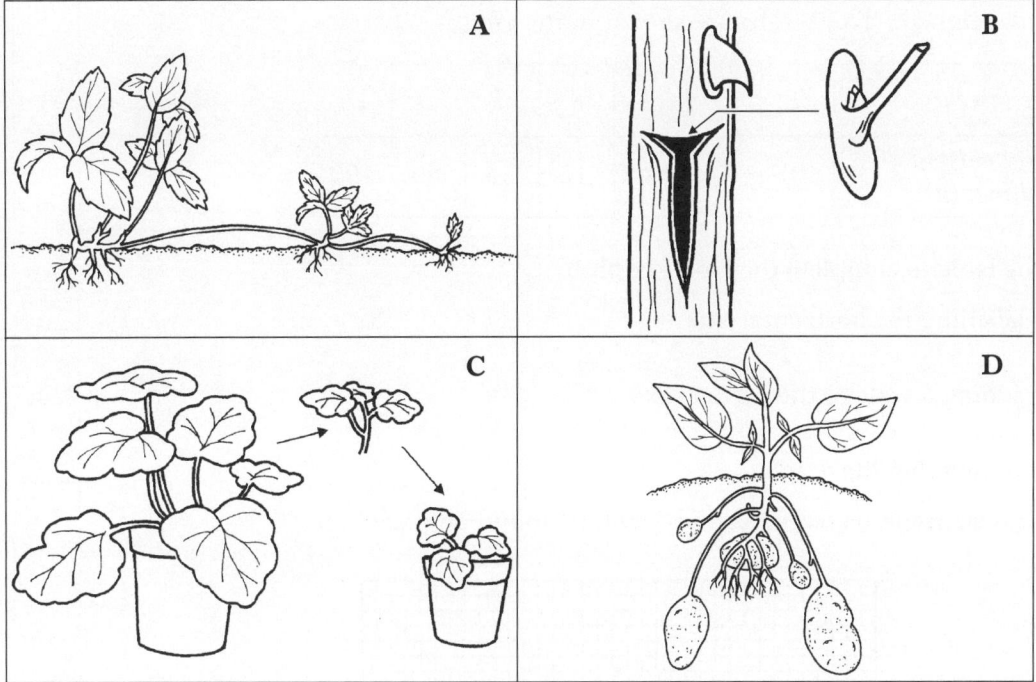

(a) Use the letters of the diagrams to identify the following.

(i) **Two** artificial methods of reproduction

letter _____ and _____   1

(ii) The diagram that shows reproduction by runners

letter _____   1

(b) Name the methods of reproduction shown by diagrams **B** and **D**.

B _____

D _____   1

[Turn over

Marks | KU | PS

8. Sweet pea seeds were planted in suitable conditions for germination and growth. Each week 20 seedlings were lifted. They were washed to clean off any soil and weighed. The results are shown in the table.

Age (weeks)	1	2	3	4	5	6	7
Total mass of 20 seedlings (g)	10	7	4	12	30	60	100

(a) Use the table to complete the **line graph** by:

   (i) labelling the horizontal axis   1

   (ii) adding a scale to the vertical axis   1

   (iii) completing the graph   1

   (Additional graph paper, if required, will be found on page 27.)

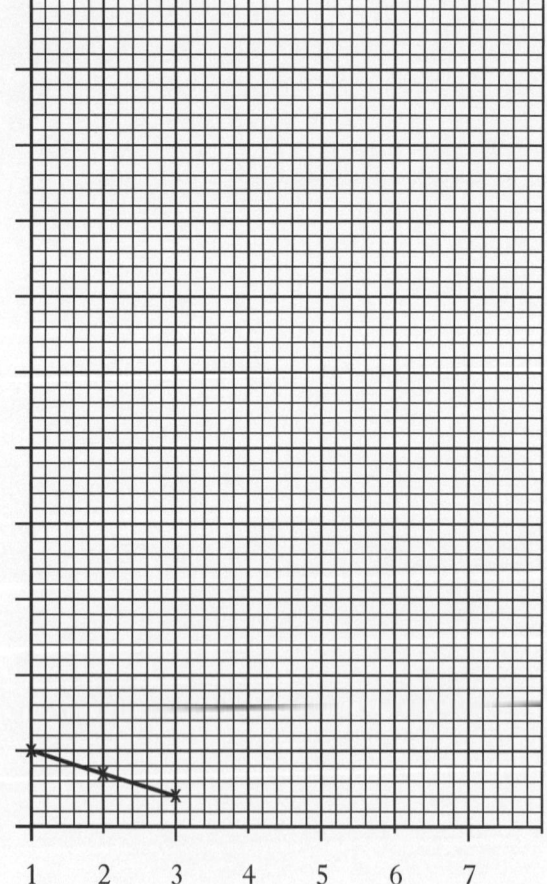

Total mass of
20 seedlings
(g)

(b) (i) At what age did the seedlings have the lowest mass?

   _____ weeks   1

8. **(b)** **(continued)**

(ii) Between which two weeks was there the greatest increase in mass?
Tick the correct box.

☐ 2 – 3

☐ 3 – 4

☐ 5 – 6

☐ 6 – 7   1

(iii) Calculate the average mass of a single seedling at age seven weeks.
Space for calculation

————— g   1

(c) (i) Name **two** factors that should be kept the same for all the seedlings during the investigation.

1 _____

2 _____   2

(ii) Weighing 20 seedlings each time reduced the error in weighing single small plants. Suggest **one** other reason for weighing 20 seedlings and calculating an average.

_____

_____   1

(iii) Removing soil from the seedlings reduced a source of error. Suggest **one** further step that should be taken before weighing the seedlings.

_____   1

(d) Predict what difference there would be in the results if the investigation was repeated in the dark.

_____

_____   1

Page thirteen   **[Turn over**

Marks | KU | PS

9. The diagram shows some of the structures found in a typical plant cell.

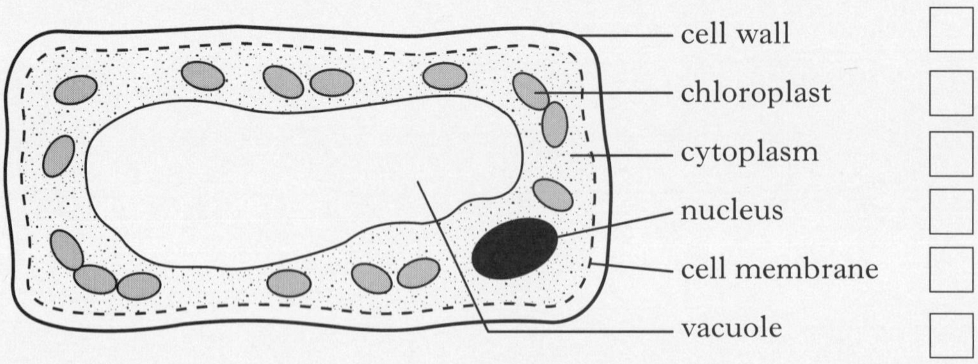

— cell wall ☐

— chloroplast ☐

— cytoplasm ☐

— nucleus ☐

— cell membrane ☐

— vacuole ☐

(a) Tick the boxes to show the structures that are also found in a typical animal cell.

2

(b) Why are cells often stained before being viewed under a microscope?

_____

1

(c) The diagram shows some plant cells as they appear when viewed under a microscope.

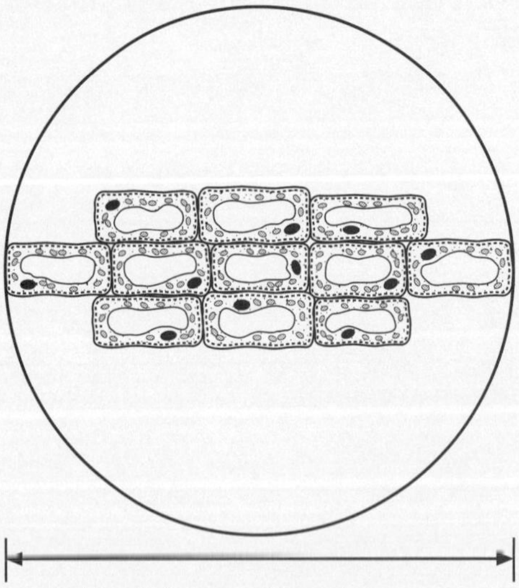

field of view = 600 micrometres

Calculate the average length of the cells.
Space for calculation

_____ micrometres

1

Marks KU PS

10. The diagram represents aerobic respiration in a cell.

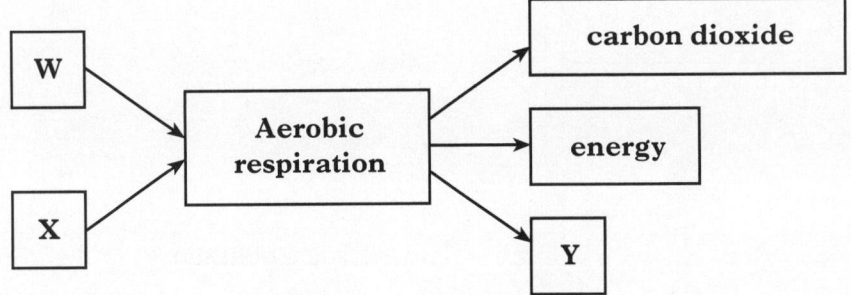

(*a*) Name the substances **W**, **X** and **Y**.

W _____

X _____

Y _____   2

(*b*) What is the source of the substance which is used in respiration and which leads to the formation of carbon dioxide?

_____   1

(*c*) The energy released during respiration can be used for chemical reactions.
Give **two** other ways in which a cell may use this energy.

1 _____

2 _____   1

[Turn over

Marks | KU | PS

11. The apparatus below was used to demonstrate diffusion.

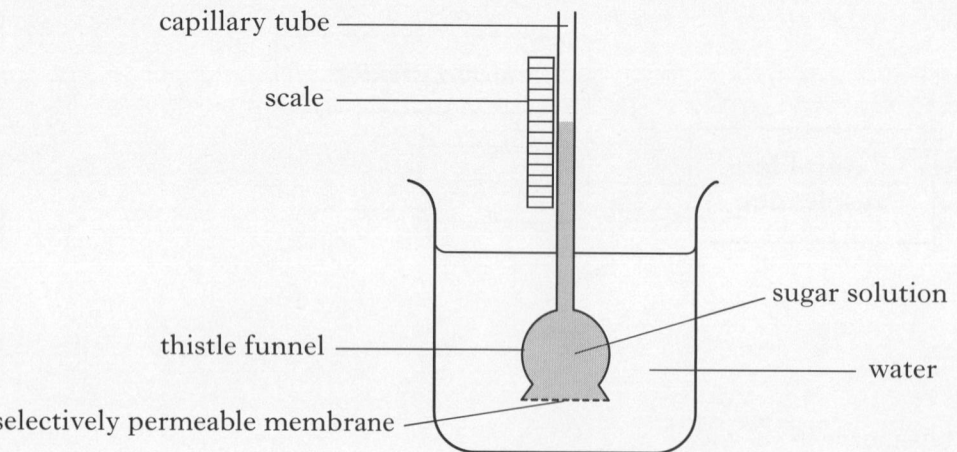

capillary tube

scale

thistle funnel

selectively permeable membrane

sugar solution

water

The height of the sugar solution in the capillary tube was measured at regular intervals. The results are shown in the graph.

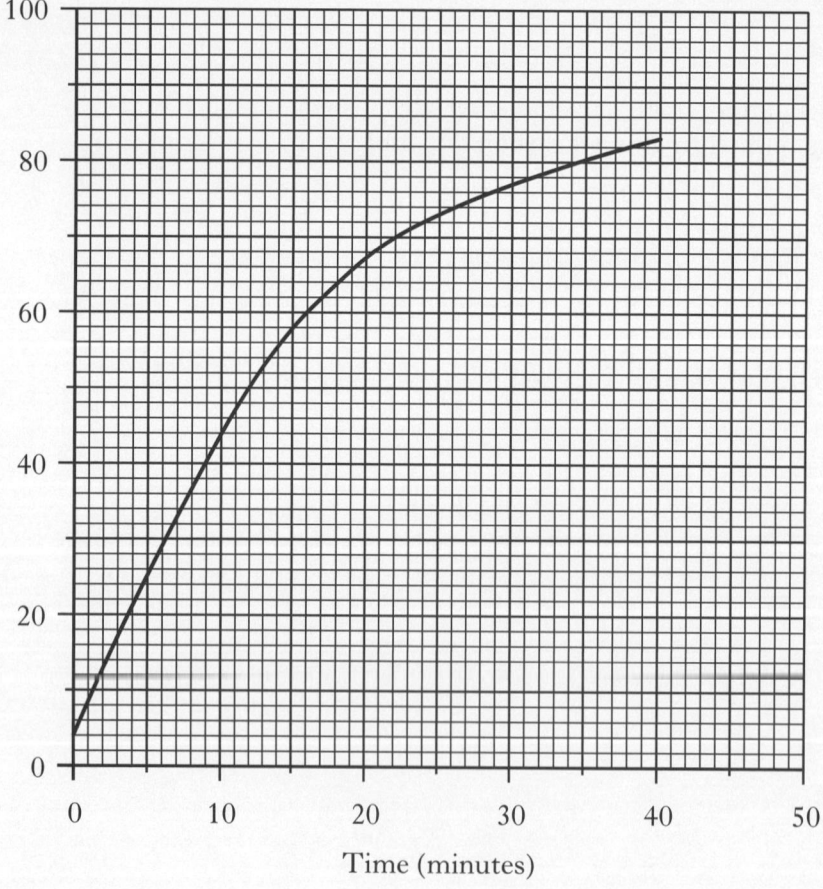

Height of sugar solution (mm)

Time (minutes)

(a) (i) What was the height of the sugar solution in the capillary tube after 10 minutes?

_____ mm

1

DO NOT WRITE IN THIS MARGIN

Marks | KU | PS

11. (*a*) (continued)

(ii) How long did it take for the sugar solution to rise from 60 mm to 70 mm?

__________ minutes **1**

(*b*) What caused the change in height of the sugar solution in the capillary tube?

Tick the correct box.

Sugar molecules moved out of the funnel. ☐

Sugar molecules moved into the funnel. ☐

Water molecules moved out of the funnel. ☐

Water molecules moved into the funnel. ☐ **1**

(*c*) Predict the height of the sugar solution in the capillary tube after 50 minutes.

__________ mm **1**

[Turn over

Marks | KU | PS

12. Read the passage below.

Adapted from *Dairy Microbiology* by the National Dairy Council.

Yoghurt is a fermented milk product that originated in the Middle East. In that part of the world it tends to be more acidic and thinner than the yoghurt that has been developed in Britain.

Yoghurt can be made from whole milk, skimmed milk, evaporated milk or dried milk. Usually a mixture of these is blended together. The milk used for yoghurt manufacture must be free of all traces of antibiotics. This is to ensure successful fermentation. The blended milk is heated to between 85 °C and 95 °C before being cooled to 32 °C. A starter culture containing bacteria is added and fermentation begins. After 12 hours, the lactic acid content reaches the desired level of between 0·8% and 1·8%.

The yoghurt is now stirred and then fruit may be added before the finished product is packaged and stored at 5 °C. The slower bacterial growth at this temperature gives the yoghurt a shelf life of approximately 10 days. After this time bacterial growth, although restricted, will increase the level of acidity to such an extent as to change the flavour and make it unacceptable to most people.

Answer the questions based on the passage.

(a) Give **two** differences between Middle Eastern yoghurt and British yoghurt.

1 _____

2 _____   1

(b) Other than whole milk, name **two** types of milk used for yoghurt manufacture.

1 _____ 2 _____   1

(c) Explain why antibiotics in the milk could prevent successful fermentation.

_____   1

(d) Name the acid produced during yoghurt production.

_____   1

DO NOT WRITE IN THIS MARGIN

Marks KU PS

12. (continued)

(*e*) What stage in yoghurt production ensures that no unwanted bacteria are present?

____ 1

(*f*) How does storage at 5 °C increase the shelf life of the yoghurt?

____ 1

(*g*) What causes the flavour of the yoghurt to change after 10 days storage?

____ 1

[Turn over

Marks KU PS

13. The diagram below shows inheritance of body colour in angelfish.

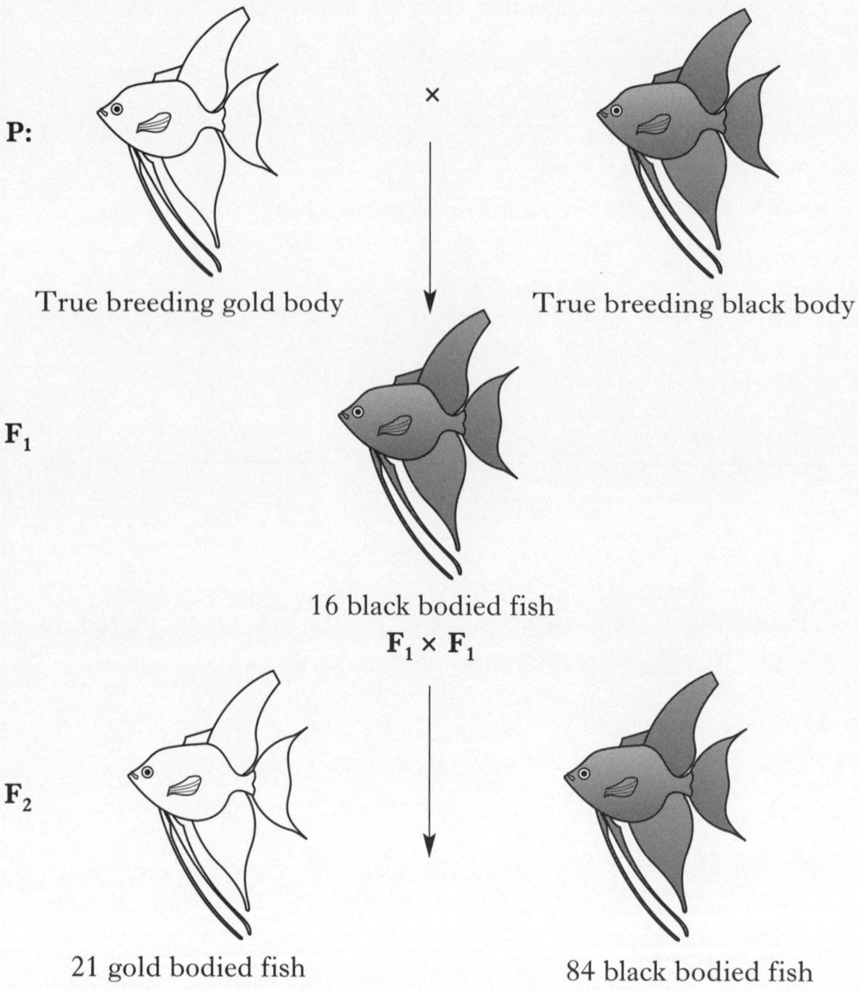
P:
×
True breeding gold body
True breeding black body
F_1
16 black bodied fish
$F_1 \times F_1$
F_2
21 gold bodied fish
84 black bodied fish

(*a*) (i) What are the **two** phenotypes used in this cross?

1 ______________________ 2 ______________________ **1**

(ii) Which characteristic is dominant?

______________________ **1**

(iii) Calculate the simple whole number ratio of **black** to **gold** bodied fish in the F_2 generation.

Space for calculation

_______ : _______

black : gold **1**

Marks | KU | PS

13. (continued)

(b) Which **one** of the following statements is true?
Tick the correct box.

The parents have the same genotypes and phenotype. ☐

All the F_1 generation have the same genotypes and phenotype. ☐

All the F_2 generation have the same genotypes and phenotype. ☐   1

(c) What type of variation is shown by the body colour of the angelfish?

_____   1

(d) Angelfish produce eggs and sperm for reproduction.
What general name is used for these sex cells?

_____   1

[Turn over

Official SQA Past Papers: General Biology 2003

DO NOT
WRITE IN
THIS
MARGIN

Marks | KU | PS

14. The skeleton provides protection for the soft organs of the body.

   (a) Give **one** other function of the skeleton.

   _____   1

   (b) The diagram shows the structure of a finger joint.

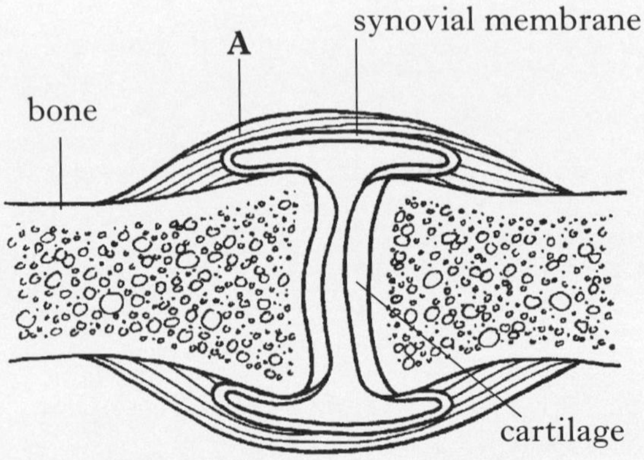

   (i) Name the part labelled **A**.

   _____   1

   (ii) What is the function of the cartilage in a joint?

   _____   1

   (c) Complete the table below about two types of moveable joints.

Range of movement allowed by the joint	Type of joint	Example
One plane		
Many planes		

2

15. The diagram shows apparatus that a pupil used to investigate gas exchange.

Air that was breathed in passed through tube A. Air that was breathed out passed through tube B.

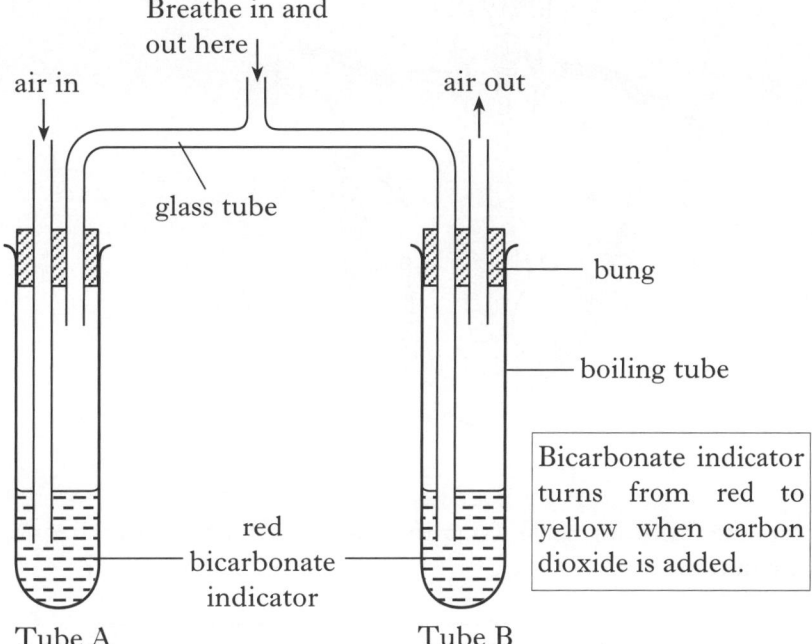

(a) (i) In which tube would the indicator change colour?

Tube _____

1

(ii) The other tube acts as a control.

What is the purpose of a control in an experiment?

_____

_____

1

(b) The experiment was repeated several times using the same apparatus.

Name **two** variables that would have to be kept constant to make sure the results were valid.

1 _____

2 _____

2

[Turn over

Official SQA Past Papers: General Biology 2003

DO NOT
WRITE IN
THIS
MARGIN

Marks | KU | PS

16. The diagram shows the chambers and blood vessels in a heart.

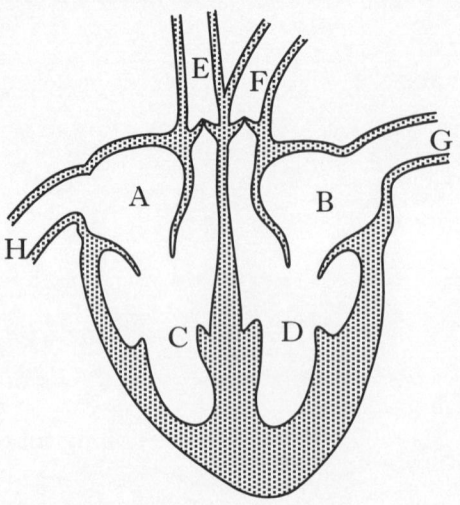

(a) Complete the following table using the correct letter from the diagram for each description.

Description	Letter
The chamber that receives blood from the body.	
The artery that carries blood from the heart to the body.	
The chamber that pumps blood to the lungs.	
The vein that carries blood from the lungs to the heart.	

3

(b) The following sentences are about blood.

Underline **one** option in each bracket to make the sentences correct.

Oxygen is carried in the blood by
$\left\{\begin{array}{l} \text{red blood cells} \\ \text{white blood cells} \\ \text{plasma} \end{array}\right\}$.

Digested food products such as glucose are carried by
$\left\{\begin{array}{l} \text{red blood cells} \\ \text{white blood cells} \\ \text{plasma} \end{array}\right\}$.

2

17. The following diagram describes some of the stages involved in transferring a gene from a **human chromosome** into a bacterial cell.

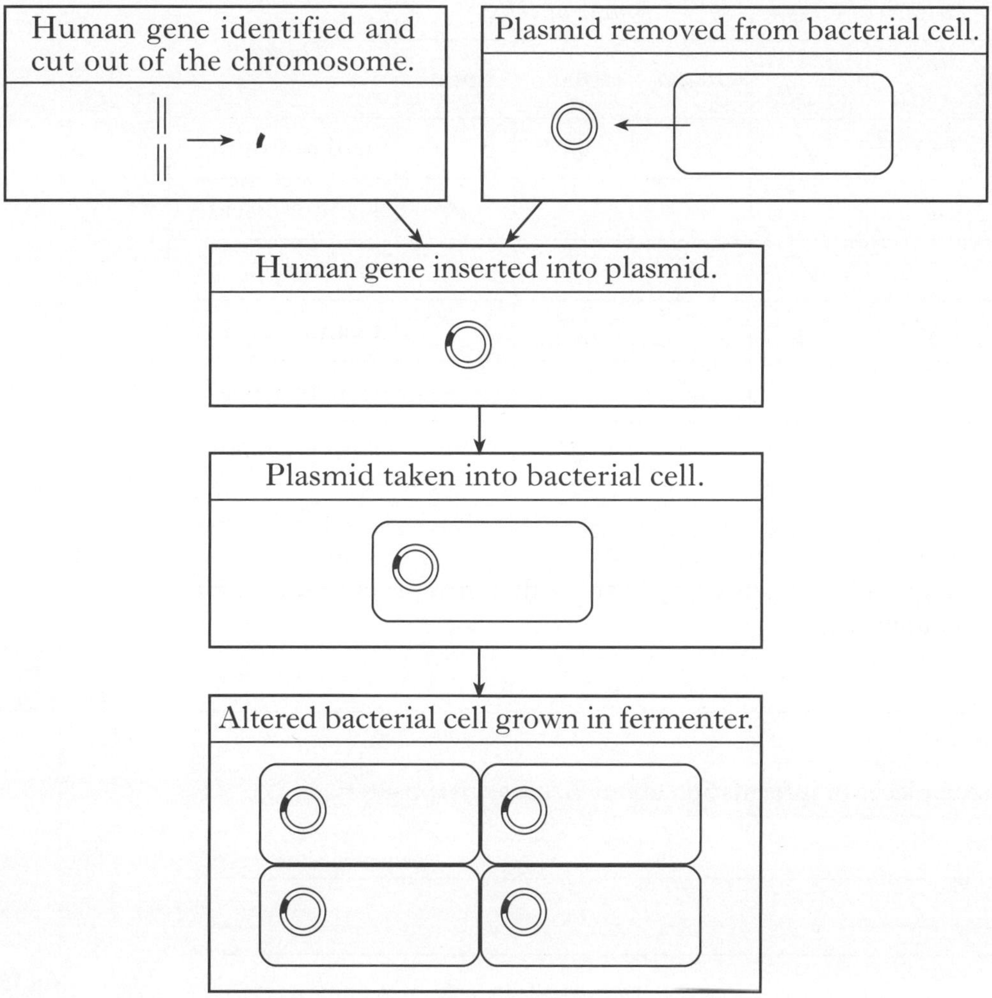

| Human gene identified and cut out of the chromosome. | Plasmid removed from bacterial cell. |

Human gene inserted into plasmid.

Plasmid taken into bacterial cell.

Altered bacterial cell grown in fermenter.

(*a*) What name is given to this procedure?

_____ 1

(*b*) Give an example of a product that can be made by bacteria as a result of this procedure. State the use of this product.

Product _____ 1

Use _____

_____ 1

(*c*) What type of reproduction is involved during the growth of the bacteria in the fermenter?

_____ 1

[**Turn over**

18. The table gives information about some disease causing bacteria.

Name of bacteria	Pattern of growth		Shape of cells			Disease
	single cells	clusters of cells	round	rod	spiral	
B. cereus		✓		✓		food poisoning
B. burgdoferi	✓				✓	Lyme's disease
S. pneumonia		✓	✓			pneumonia
C. tetani	✓			✓		tetanus
S. aureus		✓	✓			skin abscesses
E. coli	✓			✓		food poisoning

(a) Which **two** diseases are caused by bacteria that grow as clusters of cells and are round in shape?

1 _____ 2 _____   1

(b) Give **three** pieces of information about *B. burgdoferi* bacteria.

1 _____

2 _____

3 _____   1

(c) A food sample caused food poisoning.
It was found to contain rod shaped bacteria that grew as single cells.
Name the bacteria.

_____   1

[END OF QUESTION PAPER]

SPACE FOR ANSWERS
AND FOR ROUGH WORKING

ADDITIONAL GRID FOR QUESTION 5(*d*)

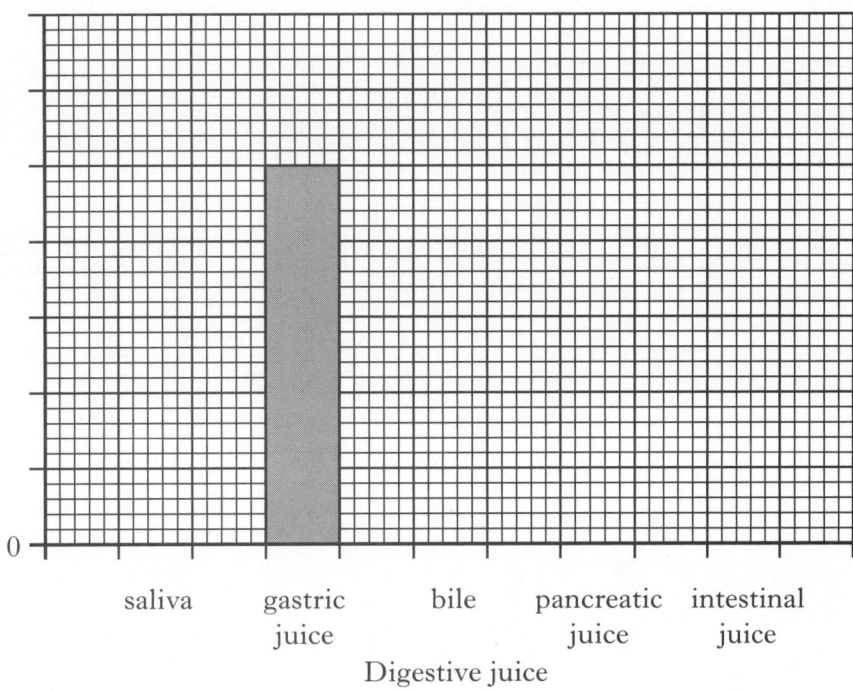

Digestive juice

ADDITIONAL GRID FOR QUESTION 8(*a*)

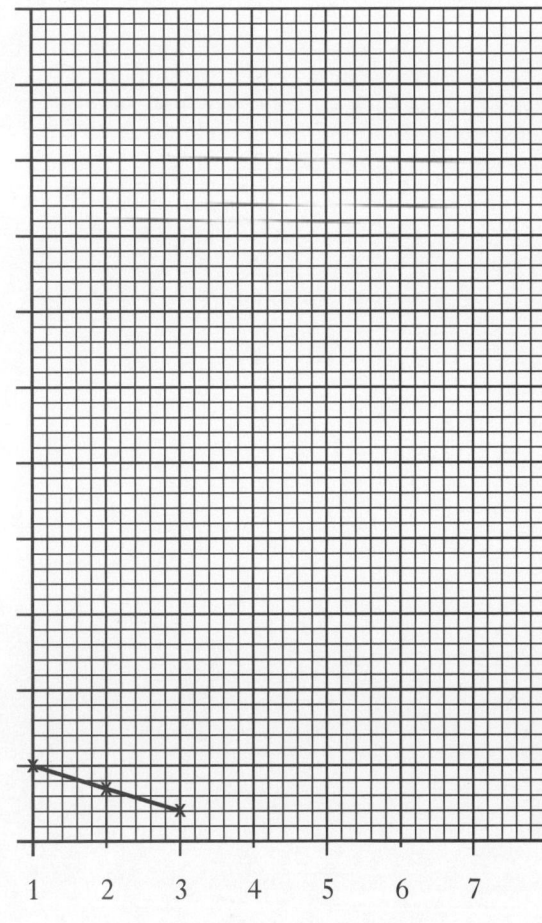

Total mass of
20 seedlings
(g)

SPACE FOR ANSWERS
AND FOR ROUGH WORKING

[BLANK PAGE]

FOR OFFICIAL USE

G

KU   PS

Total Marks

0300/401

NATIONAL
QUALIFICATIONS
2004

WEDNESDAY, 19 MAY
9.00 AM – 10.30 AM

**BIOLOGY
STANDARD GRADE**
General Level

Fill in these boxes and read what is printed below.

Full name of centre

Town

Forename(s)

Surname

Date of birth
Day Month Year   Scottish candidate number   Number of seat

1 All questions should be attempted.

2 The questions may be answered in any order but all answers are to be written in the spaces provided in this answer book, and must be written clearly and legibly in ink.

3 Rough work, if any should be necessary, as well as the fair copy, is to be written in this book. Additional spaces for answers and for rough work will be found at the end of the book. Rough work should be scored through when the fair copy has been written.

4 Before leaving the examination room you must give this book to the invigilator. If you do not, you may lose all the marks for this paper.

SCOTTISH
QUALIFICATIONS
AUTHORITY

1. (*a*) The diagram below shows a food web from a woodland ecosystem.

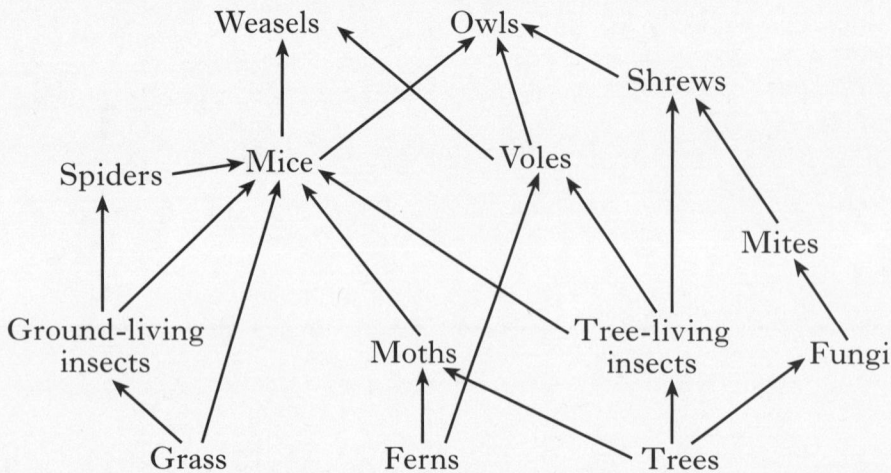

(i) Complete the table below to show each consumer from the food web and its diet.

Consumer	Diet
Mice	spiders, ground-living insects, grass, moths, tree-living insects
Moths	
	grass
Voles	
Weasels	mice, voles
Tree-living insects	trees
	tree-living insects, mites
Fungi	trees
Mites	
	ground-living insects
Owls	

3

DO NOT WRITE IN THIS MARGIN

Marks | KU | PS

1. (*a*) (continued)

(ii) Use the food web to complete the food chain below, consisting of four organisms.

ferns → ______ → ______ → ______ **1**

(*b*) Trees are producers and mice are consumers.

What is the meaning of the terms producer and consumer?

Producer ____

____ **1**

Consumer ____

____ **1**

[Turn over

Marks | KU | PS

2. Some features of six species of the buttercup family are shown in the table below.

Species name	Leaves	Runners	Stem
Greater spearwort	toothed	present	hairy
Meadow buttercup	lobed	absent	hairy
Lesser celandine	heart-shaped	absent	hairless
Creeping buttercup	lobed	present	hairy
Lesser spearwort	toothed	absent	hairless
Celery-leaved buttercup	lobed	absent	hairless

(a) Use the information in the table to complete the key below.

Write the correct feature on each dotted line and the correct names in the empty boxes.

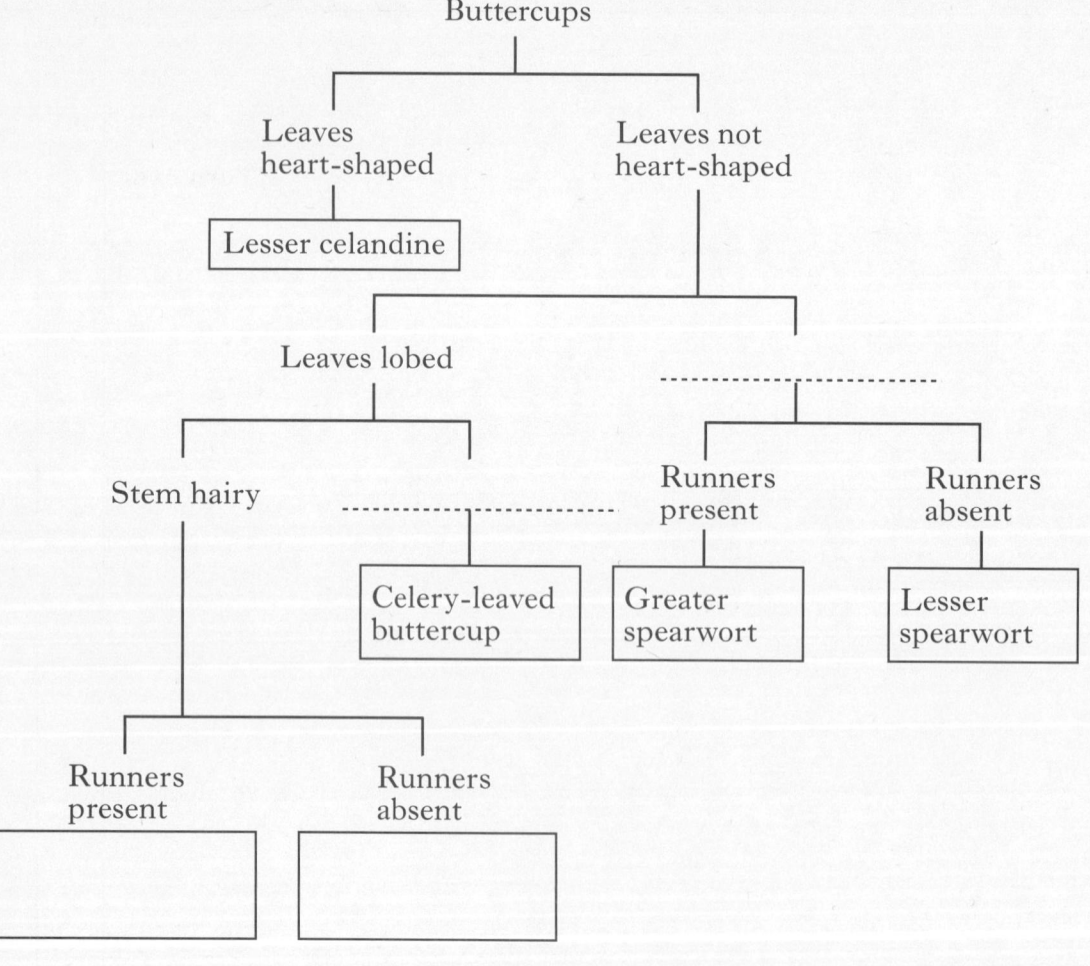

3

DO NOT
WRITE IN
THIS
MARGIN

Marks KU PS

2. (continued)

(*b*) Which feature could be used to distinguish between a Lesser celandine and a Lesser spearwort?

_____

1

(*c*) Which features do the Meadow buttercup and the Celery-leaved buttercup have in common?

_____

1

[Turn over

Marks KU PS

3. (*a*) A population survey of barnacles and mussels between the high and low tide marks of a rocky shore was carried out using quadrats.

The results are shown in the table below.

Tide mark	High								→	Low
Quadrat number	1	2	3	4	5	6	7	8	9	10
Number of mussels	0	2	15	31	32	34	50	55	58	60
Number of barnacles	52	51	37	40	40	23	15	17	15	10

   (i) On the grid below, complete the bar chart by

   1. adding a scale to the vertical axis   **1**

   2. plotting the bars for the barnacles in quadrats 5–10   **1**

   (An additional grid, if needed, will be found on page 27.)

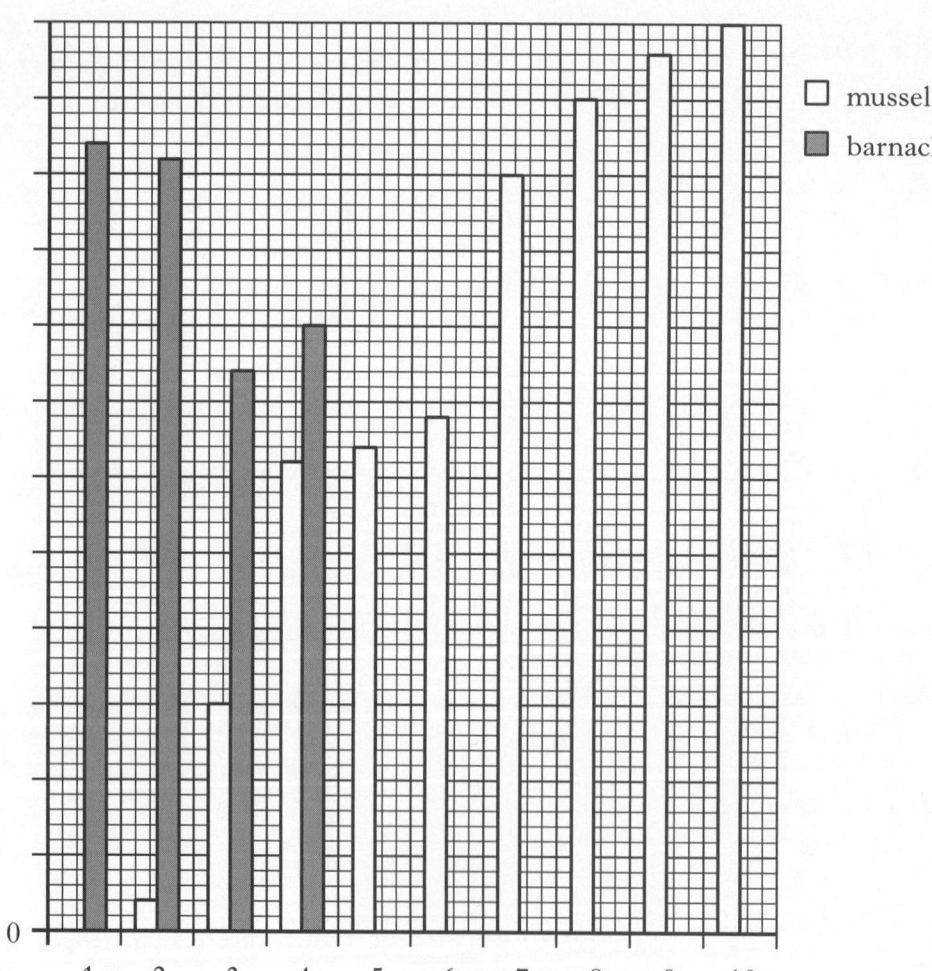

Number of organisms

Quadrat number

3. (a) (continued)

(ii) Calculate the average number of barnacles per quadrat.
Space for calculation

Average number _____

1

(iii) What is the trend shown by the number of mussels from the high to the low tide marks?

_____

1

(b) The mussels and the barnacles are in competition with each other.
State **one** possible effect on the mussel population of **reduced competition** from barnacles.

_____

_____

1

(c) The following factors affect populations of barnacles and mussels.
Underline **two** abiotic factors from the list.

List of factors   water temperature

disease

predators

salt concentration

food supply

1

(d) A rocky shore ecosystem consists of a community of organisms and one other part.
Name the other part.

_____

1

[Turn over

DO NOT WRITE IN THIS MARGIN

Marks | KU | PS

4. (*a*) In an investigation on photosynthesis, two bell jars were set up as shown below and left in bright light.

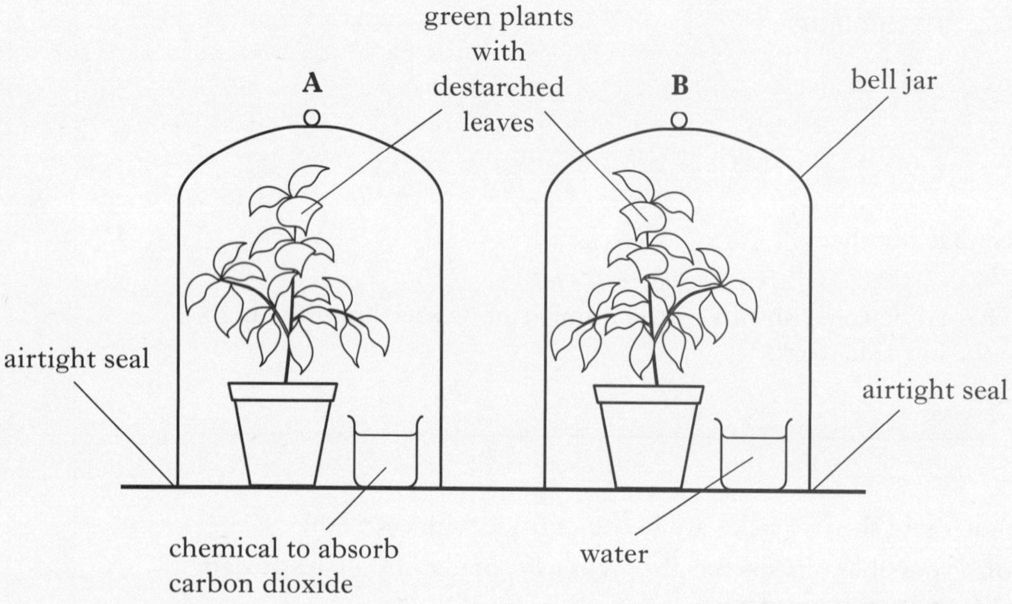

After 48 hours a leaf was removed from each plant and tested for starch.

(i) In which plant would photosynthesis take place? Give a reason for your answer.

Plant ____________

Reason ______________________________

______________________________ **1**

(ii) Name a product of photosynthesis, other than carbohydrate.

______________________________ **1**

(iii) Why were the plants destarched before being used in the investigation?

____

______________________________ **1**

(iv) Give **one** feature of the plants that would have to be kept the same to allow a fair comparison in the investigation.

______________________________ **1**

4. **(continued)**

(b)   Name the structures in a leaf through which gases can pass.

_____

1

(c)   Name the chemical found in leaves that converts light energy into chemical energy during photosynthesis.

_____

1

(d)   The grid below refers to parts of a flower.

A sepal	B petal	C stamen	D anther
E stigma	F ovary	G nectary	H ovule

Use letters from the grid to answer the following questions.

(i)   Which structure protects the flower bud?

_____

1

(ii)   Which structure receives pollen grains?

_____

1

(iii)   Which structure develops into a fruit after fertilisation?

_____

1

[Turn over

DO NOT
WRITE IN
THIS
MARGIN

Marks | KU | PS

5. (*a*) The diagram represents the reproductive system of a human female.

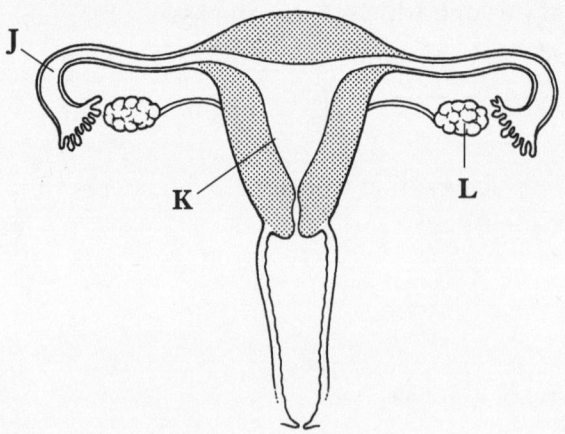

(i) Name the parts labelled on the diagram.

J _____

K _____

L _____

2

(ii) In the table below, match each letter from the diagram to its correct function.

Function	Letter
Eggs produced	
Fertilisation takes place	
Fertilised egg becomes attached	

2

(*b*) Tick (✓) boxes in the table to indicate whether each of the following statements is true for eggs, sperm, or both.

Statement	Eggs	Sperm
Contain a food store for developing fetus		
Swim using a tail		
Produced in testes		
In most fish, are deposited into the water		
Are gametes		

2

Marks | KU | PS

6. The apparatus shown was set up to investigate the behaviour of woodlice.

gauze platform   woodlice

Side A   **Side B**

drying agent   water

At the start of the investigation 20 woodlice were placed in the centre of the chamber. After 10 minutes there were 2 on side A and 18 on side B.

(a) What environmental factor was being investigated?

_____   1

(b) Describe the response of the woodlice in the investigation.

_____

_____   1

(c) Why were the woodlice left for ten minutes before the results were taken?

_____

_____   1

(d) Why were 20 woodlice used, rather than one?

_____   1

(e) Name **one** abiotic factor which should be kept constant during the investigation.

_____   1

(f) Suggest **two** changes which could be made to the apparatus in order to investigate the response of woodlice to light.

1 _____

_____   1

2 _____

_____   1

Marks | KU | PS

7. (a) Complete the table by using all the letters from the list to identify the parts found in each type of cell.

Each part may be used **once** or **more than once**.

Parts of cells

A   cell membrane
B   cell wall
C   chloroplast
D   cytoplasm
E   nucleus

Leaf cell	*Cheek cell*

2

(b) Use the information in the table below to answer the questions about liquids used in preparing microscope slides.

Type of cell	*Liquid used*	*Effect*
human cheek cell	methylene blue	nucleus turns blue
onion epidermal cell	iodine solution	nucleus turns yellow
human skin cell	eosin	cytoplasm turns pink
onion root cell	acetic orcein	chromosomes turn red

(i) Name **two** liquids used to prepare plant cells.

1 _____

2 _____

1

(ii) What effect does eosin have on skin cells?

_____

1

(iii) Which liquid could be used to show stages of mitosis?

_____

1

7. (continued)

(c) What name is given to a liquid that is used to make the parts of a cell clearer when viewed under a microscope?

_____

1

(d) The magnification of a microscope is calculated using the following formula.

Total magnification = eyepiece lens × objective lens
magnification   magnification

Use the formula to complete the following table.

The same eyepiece was used each time.

Power	Eyepiece lens magnification	Objective lens magnification	Total magnification
Low	× 12	× 4	
Medium		× 10	
High	× 12		× 480

2

[Turn over

Marks | KU | PS

8. (a) The statements in the table describe the movement of substances into or out of cells.

Number	Statement
1	glucose moves from the small intestine into the blood
2	water enters root cells from the soil
3	carbon dioxide passes from the blood into the lungs

(i) Which statement is an example of osmosis?

Statement number _____

1

(ii) What term could be used to describe the movement of substances in all of the examples?

_____

1

(b) Pieces of potato were weighed, placed in sugar solutions of different concentrations for one hour, then reweighed.

The graph below shows the percentage change in mass at each concentration.

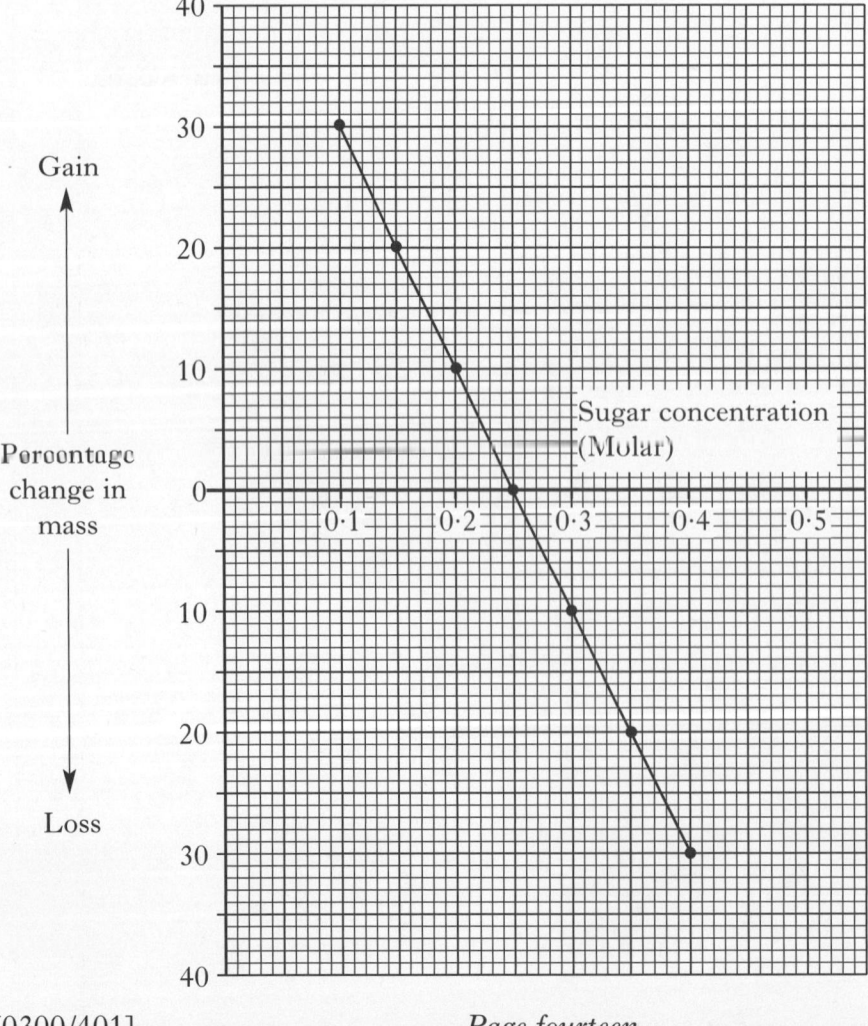

DO NOT WRITE IN THIS MARGIN

Marks KU PS

8. (*b*) (continued)

(i) The movement of what substance is responsible for the change in mass?

__________________________ **1**

(ii) What was the percentage change in mass of the piece of potato placed in the 0·15 Molar solution?

_________ % **1**

(iii) What was the concentration of the solution which caused the potato to lose 30% of its original mass?

_________ Molar **1**

(iv) At what concentration was there no change in mass of the potato?

_________ Molar **1**

[Turn over

9. Read the following passage carefully.

Adapted from "*Stirring Stuff's in the bag*", The Herald, April 2002.

Pausing for a cup of tea is a good way to take time out in a busy day. About 135 million cups are consumed in Britain daily.

Favourite "cuppas" include first thing in the morning before getting ready for work, during a busy day and at the end of the day to relax. Relaxation is the most common mood when taking a tea break.

As well as relieving stress, tea can also be a life-saver. Research has shown that the great British "cuppa" has disease-fighting capabilities. A cup of tea can have protective effects against cancer and heart disease. A mixture of green tea and black tea rubbed on cancerous areas reduced cell growth. Tests show that tea slows the development of lung cancers and some bowel cancers. It is also thought to decrease the risk of cancer of the digestive system. Red tea from South Africa is rich in antioxidants and free from tannin and caffeine which are found in many other teas.

The three basic types of tea, black, green and oolong, give rise to more than 3000 varieties, each having its own distinct character. People are now trying different styles of teas such as organic, Chai spice, decaffeinated, herbal and iced tea.

Answer the following questions, based on the above passage.

(*a*) How much tea is drunk in Britain daily?

_____

1

(*b*) What is the most common mood whilst drinking tea?

_____

1

(*c*) Apart from cancer, what disease can tea help prevent?

_____

1

(*d*) Name **three** types of cancer that tea may help prevent.

1 _____ 2 _____ 3 _____

1

(*e*) What **two** substances are not present in South African red tea?

_____

1

(*f*) Name **three** styles of tea, mentioned in the passage, that people are now trying.

1 _____ 2 _____ 3 _____

1

Marks KU PS

10. The bar graph shows the body lengths in a population of 300 budgerigars. The pie chart shows the colours in the same population.

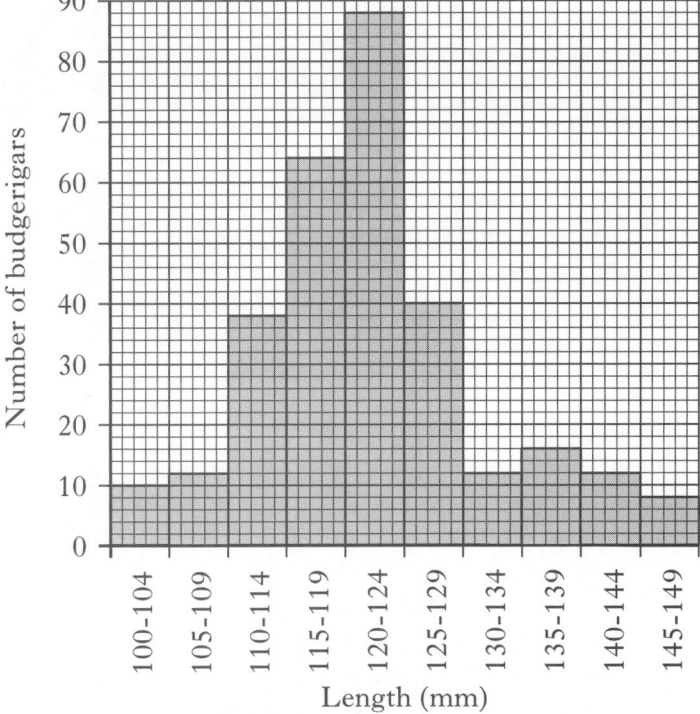

Colour of budgerigars

(a) How many budgerigars are in the range 110 to 119 mm long?
Space for calculation

_____ 1

(b) Which of the two characteristics is an example of discontinuous variation?

_____ 1

(c) What percentage of the budgerigars are blue?
Space for calculation

_____ % 1

[Turn over

Marks | KU | PS

11. (*a*) An investigation was carried out into the growth of a bacterial culture. The numbers of bacteria were counted every 30 minutes and the results are shown in the table below.

Time (minutes)	0	30	60	90	120	150
Number of bacteria (thousands per mm³)	3	6	12	24	48	96

(i) What happens to the number of bacteria every 30 minutes?

_____

1

(ii) Complete the line graph below by

1   adding a suitable scale to the y-axis

1

2   adding a label to the x-axis

1

3   plotting the graph.

1

(An additional grid, if needed, will be found on page 28.)

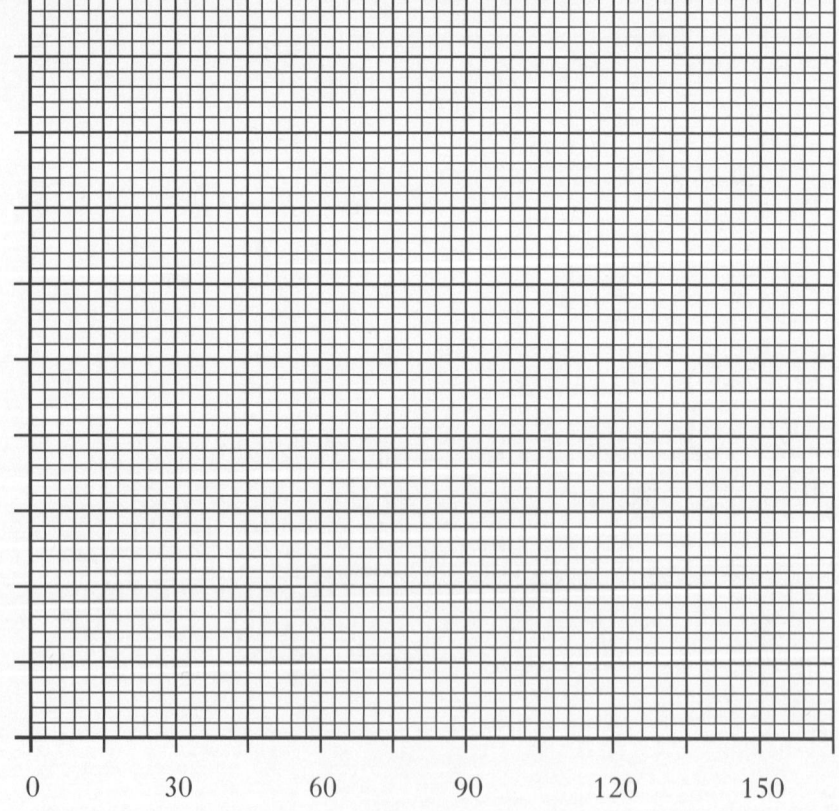

Number of
bacteria
(thousands
per mm³)

0   30   60   90   120   150

(iii) Assuming no change in conditions, how many bacteria cells would be present after 240 minutes?
Space for calculation

_____ thousands per mm³

1

Marks | KU | PS

11. (continued)

(b) The following diagrams show four stages of mitotic cell division but not in the correct order.

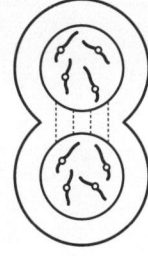 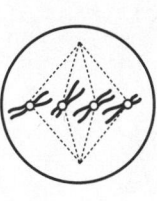  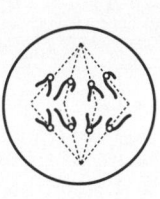

A   B   C   D

Arrange the letters from the diagrams to put the stages into the correct order. The first stage has been given.

1st stage   C

2nd stage   _____

3rd stage   _____

4th stage   _____   1

(c) Complete the following sentence by underlining the correct option in each group.

In comparison with the original cell, the number of chromosomes present in a cell produced by mitosis is { greater / smaller / the same } and it contains { different / the same } information.   1

[Turn over

Official SQA Past Papers: General Biology 2004

DO NOT
WRITE IN
THIS
MARGIN

Marks | KU | PS

12. (a) The diagram shows some of the structures of the human eye.

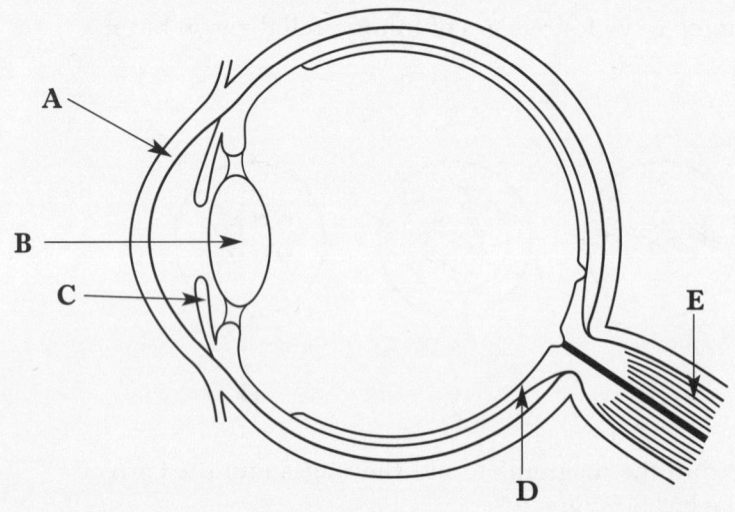

Complete the table to show the names and functions of the structures labelled.

Letter	Name of structure	Function
A		Allows light to enter the eye
B		
C	Iris	
D		Converts light into electrical impulses
E	Optic nerve	

3

(b) Humans have two eyes and two ears. What does this contribute to their sight and hearing?

Sight _____

1

_____

Hearing _____

1

_____

Marks | KU | PS

12. (continued)

(*c*) The diagram represents the flow of information in the human nervous system.

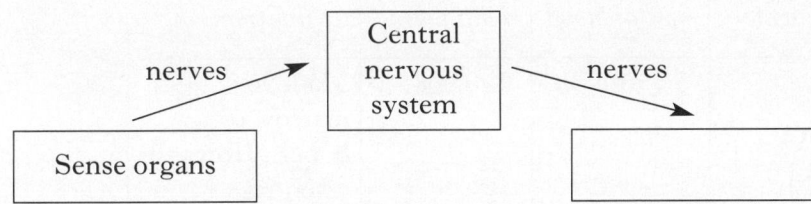

(i) Complete the diagram by writing the missing word in the box. **1**

(ii) Name the two main parts of the central nervous system.

1 _____ 2 _____ **1**

[Turn over

Page twenty-one

Marks | KU | PS

13. (a) The table gives information about components of the blood.
Use the information provided to answer the questions which follow.

Appearance under a microscope (not drawn to the same scale)	Number per mm³ of blood	Diameter in millimetres	Additional information
Red blood cells	5·5 million	0·008	Made in marrow of bones. Iron essential. 2 million made each second. Last for about 4 months.
White blood cells	8000	0·02	Made in marrow of bones or in lymph nodes. Fight infection by engulfing bacteria or producing antibodies.
Platelets	400,000	0·003	Made in marrow. Contain proteins which form blood clots.

(i) Name **two** places where blood cells are made.

1 _____ 2 _____   1

(ii) Which cells are the largest?

_____   1

(iii) Which component is present in the greatest numbers?

_____   1

(iv) What type of substance is needed to form blood clots?

_____   1

DO NOT
WRITE IN
THIS
MARGIN

Marks | KU | PS

13. (*a*) (**continued**)

(v) Describe **two** ways in which white blood cells fight infection.

1 _____

2 _____   **1**

(vi) On average, how many red blood cells are made in an hour?
Space for calculation

_____ million   **1**

(*b*) The diagram below represents the site of gas exchange between a blood vessel and the muscle cells of a mammal.

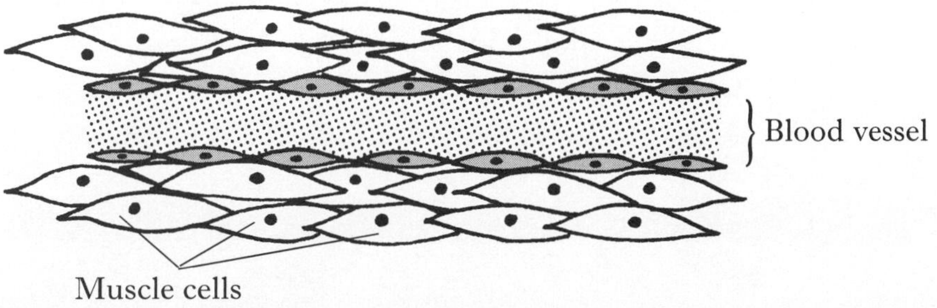

} Blood vessel

Muscle cells

(i) Name the type of blood vessel shown.

_____   **1**

(ii) On the diagram, write the letter **H** to indicate an area where the oxygen concentration is relatively high and the letter **L** to indicate where it is relatively low.   **1**

(*c*) In which component of blood is most of the oxygen carried?

_____   **1**

[Turn over

Marks | KU | PS

14. (*a*) A mule is produced by mating a horse and a donkey.
Mules are always infertile. What information does this provide about horses and donkeys?

_____

1

(*b*) The diagram below shows inheritance of colour in onions.

 ×

Generation A   Red   White

Generation B   all Red
                    Generation B onions self-crossed

Generation C   36 Red   9 White

(i) Which onion colour is dominant?

_____

1

(ii) Complete the table with the correct symbols to identify each of the generations shown in the diagram.

Generation	Symbol
A	**P**
B	
C	

1

(iii) Calculate the simple whole number ratio of red onions to white onions produced in Generation C.

Space for calculation

_____ : _____
Red onions   White onions

1

[0300/401]   *Page twenty-four*

DO NOT
WRITE IN
THIS
MARGIN

Marks | KU | PS

15. Thalassaemia is an inherited disease which prevents people producing blood cells. The family tree shows inheritance of thalassaemia.

☐ Unaffected male   ■ Thalassaemic male

◯ Unaffected female   ● Thalassaemic female

(a)   (i)   Which of the following statements about Parents A and B is true?
Tick (✓) the correct box.

Both have the thalassaemic gene.   ☐

One has the thalassaemic gene.   ☐

Neither has the thalassaemic gene.   ☐   1

(ii)   Give a reason for your answer.

_____   1

(b)   What proportion of the children of Parents A and B were thalassaemic?

_____   1

(c)   Doctors can test for thalassaemia by examining the cells of a fetus. The cells are obtained by inserting a needle into the mother's uterus and withdrawing fluid from around the fetus.

What name is given to this procedure?

_____   1

[Turn over

Official SQA Past Papers: General Biology 2004

DO NOT
WRITE IN
THIS
MARGIN

Marks | KU | PS

16. Yeast is a micro-organism which carries out fermentation.

(a) Complete the following word equation for fermentation in yeast.

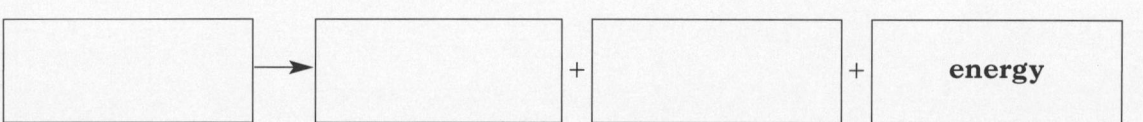

energy

1

(b) Name **two** manufacturing processes which depend on fermentation by yeast.

1 _____

2 _____   **1**

(c) Complete the following sentence by underlining the correct word in each group.

Yeast is a $\left\{ \begin{array}{c} \text{fungus} \\ \text{bacterium} \end{array} \right\}$ and is $\left\{ \begin{array}{c} \text{single-} \\ \text{multi-} \end{array} \right\}$ celled.   **1**

(d) Describe the precautions which should be taken with each of the following items when working with micro-organisms.

1 Bench surfaces _____

_____   **1**

2 Wire loops for inoculating a plate _____

_____   **1**

(e) Petri dishes half-filled with agar gel are used to grow micro-organisms.

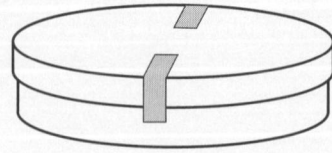

Explain why Petri dishes containing micro-organisms must be kept closed.

_____   **1**

[END OF QUESTION PAPER]

SPACE FOR ANSWERS
AND FOR ROUGH WORKING

ADDITIONAL GRID FOR QUESTION 3(a)(i)

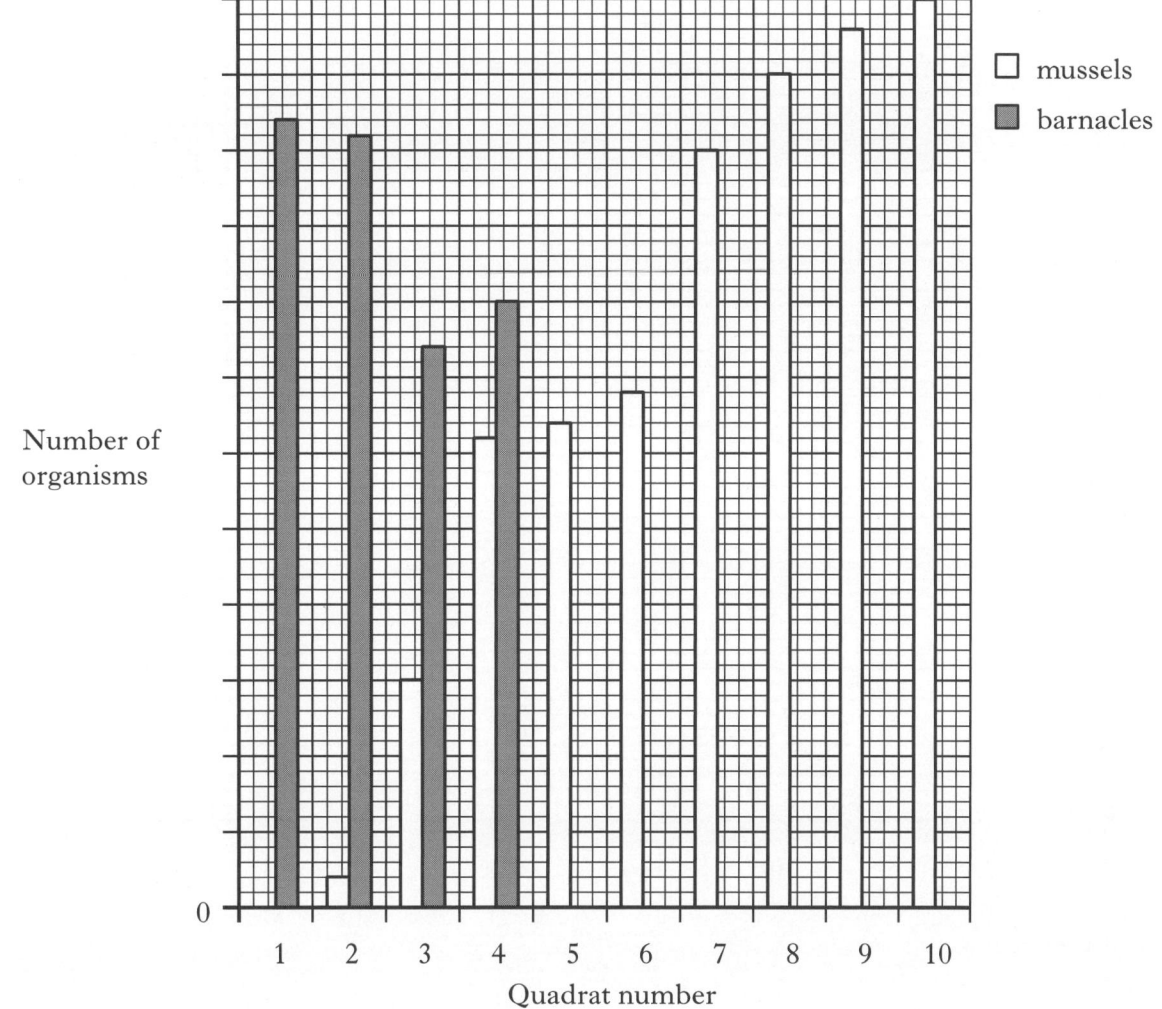

SPACE FOR ANSWERS
AND FOR ROUGH WORKING

ADDITIONAL GRID FOR QUESTION 11(*a*)(ii)

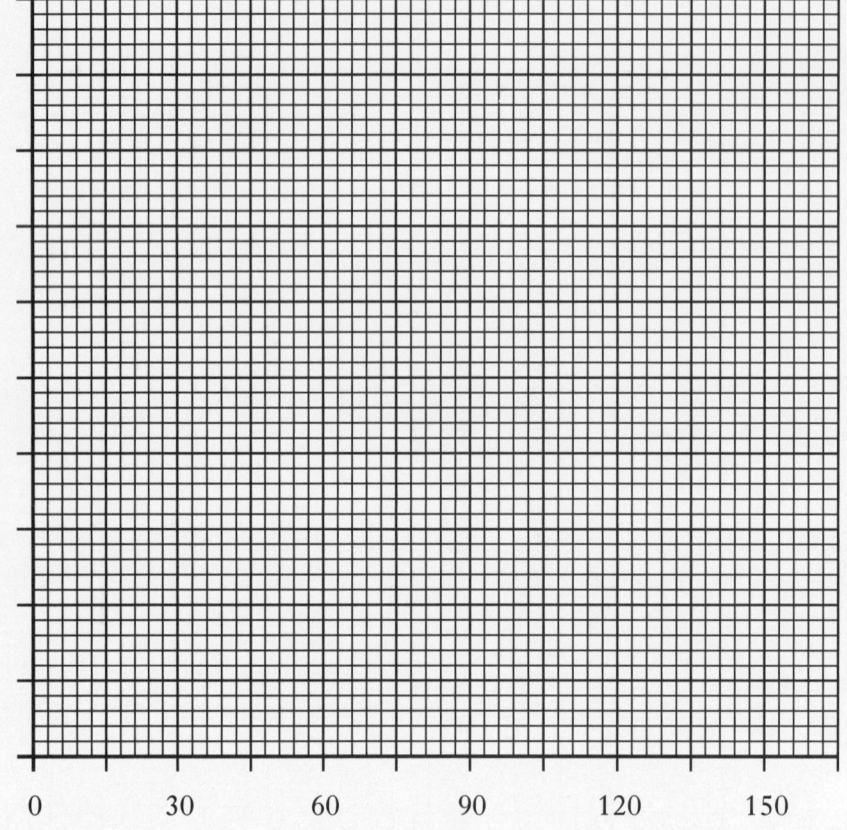

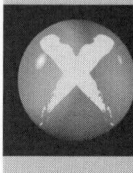

[BLANK PAGE]

FOR OFFICIAL USE

G

KU	PS

Total Marks

0300/401

NATIONAL
QUALIFICATIONS
2005

WEDNESDAY, 18 MAY
9.00 AM – 10.30 AM

BIOLOGY
STANDARD GRADE
General Level

Fill in these boxes and read what is printed below.

Full name of centre

Town

Forename(s)

Surname

Date of birth
Day Month Year   Scottish candidate number   Number of seat

1 All questions should be attempted.

2 The questions may be answered in any order but all answers are to be written in the spaces provided in this answer book, and must be written clearly and legibly in ink.

3 Rough work, if any should be necessary, as well as the fair copy, is to be written in this book. Additional spaces for answers and for rough work will be found at the end of the book. Rough work should be scored through when the fair copy has been written.

4 Before leaving the examination room you must give this book to the invigilator. If you do not, you may lose all the marks for this paper.

SCOTTISH
QUALIFICATIONS
AUTHORITY

©

1. Part of a woodland food web is shown below.

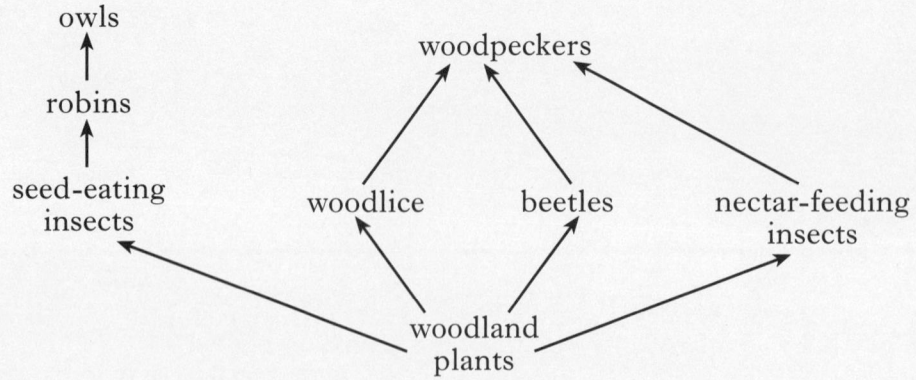

(a)　(i)　What do the arrows in the food web represent?

_____

1

(ii)　How many food chains in the food web involve woodpeckers?

_____

1

(iii)　Blue tits eat woodlice. Hawks eat blue tits and robins.

Add blue tits and hawks to the food web diagram to show their feeding relationships.

1

(b)　A study of the populations of beetles and woodlice in an area of woodland was carried out over a number of years.

The results are shown below.

Year of study	Number of beetles	Number of woodlice
1	563	540
2	641	672
3	682	698
4	117	940

(i)　Use information from **both** the table **and** the food web to suggest an explanation for the drop in the number of beetles in year 4 of the study.

_____

_____

1

1.(b) (continued)

(ii) During the same period, the numbers of owls increased.

Explain this change in terms of their birth rate and death rate.

_____

_____   1

(iii) What change in the population of the robins could have caused the increase in the number of owls?

_____   1

(c) When a plant or animal dies, decay takes place.

Choose words from the box below to complete the following sentences about decay.

You may use each word **once**, **more than once** or **not at all**.

soil	animals	nutrients
plants	protein	micro-organisms

Decay is carried out by _____ .

This process releases _____ which can be absorbed

by _____ .   2

[Turn over

Marks | KU | PS

2. (a) The following table gives the results of an investigation on the factors affecting seed germination.

Test tube	Conditions provided		
	Temperature (°C)	Water present	Oxygen present
1	20	yes	yes
2	20	yes	no
3	20	no	yes
4	0	yes	yes
5	0	no	yes

In which tube(s) would germination occur?

_____

1

(b) The diagram shows a section through a seed.

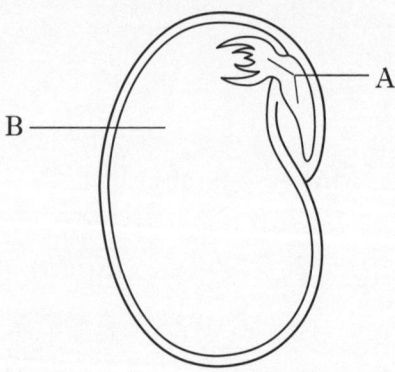

(i) Name the parts labelled A and B.

A _____

1

B _____

1

(ii) Name the part of a seed which protects the internal structures.

_____

1

DO NOT
WRITE IN
THIS
MARGIN

Marks | KU | PS

2. (continued)

(c) The table below shows the time from germination to flowering for some plant species.

Plant species	Time from germination to flowering (years)
Rock rose	2·0
Hollyhock	0·5
Broom	3·0
Birch	10·0
Phlox	1·0
Berberis	5·5

Use the information from the table to complete the **bar chart** below by:

(i) labelling the vertical axis;   1

(ii) adding an appropriate scale to the vertical axis;   1

(iii) drawing the bars.   1

(Additional graph paper, if required, will be found on page 30.)

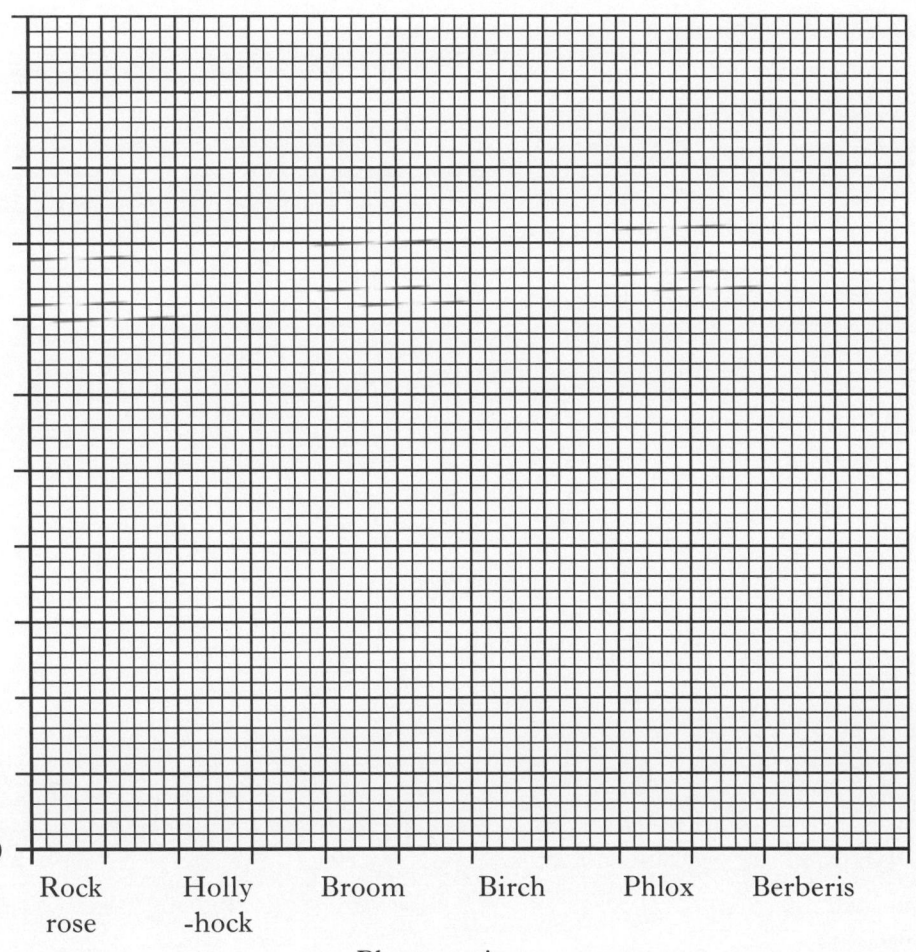

Plant species

[Turn over

3. The table below gives information about the harvest of softwood timber over a 3 year period.

Country	Timber harvested each year (m³)		
	1999	2000	2001
Scotland	2292	2496	2883
England	699	881	913
Wales	385	463	663
Total		3840	

(a) Complete the table to show the totals harvested in 1999 and in 2001.

Space for calculations

(b) What was the increase in timber harvested in Wales between 1999 and 2001?

Space for calculation

_____ m³

(c) What percentage of the total timber harvested in 2000 was produced in Scotland?

Space for calculation

_____%

Marks | KU | PS

4. The following sentences refer to food manufacture and transport in plants.
Underline **one** alternative in each bracket to make the sentences correct.

(a) Light energy is converted to chemical energy by $\left\{ \begin{array}{l} \text{carbon dioxide} \\ \text{chlorophyll} \end{array} \right\}$.   **1**

(b) Food is transported from the leaves in $\left\{ \begin{array}{l} \text{xylem} \\ \text{phloem} \end{array} \right\}$.   **1**

(c) The raw materials for photosynthesis are $\left\{ \begin{array}{l} \text{carbon dioxide} \\ \text{oxygen} \end{array} \right\}$ and

$\left\{ \begin{array}{l} \text{glucose} \\ \text{water} \end{array} \right\}$.   **1**

(d) Food may be stored in the leaves as $\left\{ \begin{array}{l} \text{glucose} \\ \text{starch} \end{array} \right\}$.   **1**

[Turn over

Marks | KU | PS

5. The diagram shows a choice chamber which could be used to investigate the behaviour of woodlice.

woodlice

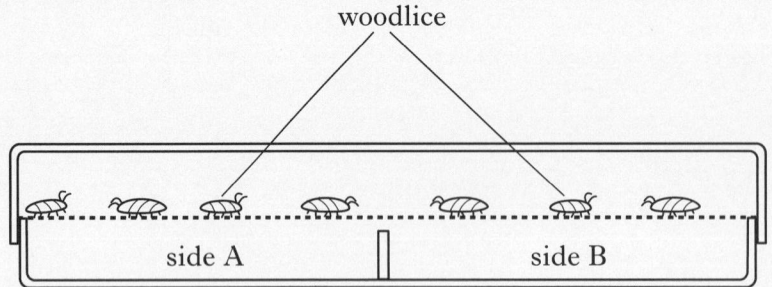

side A   side B

(*a*) How could the choice chamber be set up to study the effect of light on the behaviour of woodlice?

_____

_____ 1

(*b*) (i) Name **one** other abiotic factor which may affect woodlice behaviour that can be investigated using a choice chamber.

_____ 1

(ii) How could the choice chamber be set up to investigate this other abiotic factor?

_____

_____ 1

6. The effect of practice on the reaction times of three volunteers was investigated. A buzzer was sounded and the time taken to stop a clock was measured.

   Each volunteer was tested 10 times.

   The results are shown in the table.

Volunteer \ Attempt	Reaction time (milliseconds)									
	1	2	3	4	5	6	7	8	9	10
A	256	250	210	207	201	192	187	164	162	154
B	234	227	218	201	200	185	179	161	153	147
C	218	200	195	192	186	178	160	149	136	131

(a) Why were three volunteers tested rather than one?

_____   1

(b) The average reaction time of the three volunteers' first attempts was 236 milliseconds.

   Calculate the average reaction time of their final attempts.
   Space for calculation

   _____ milliseconds   1

(c) From the results of the investigation, describe the effect of practice on reaction time.

_____

_____   1

[Turn over

Marks | KU | PS

7. (*a*) The following list contains descriptions of stages in the reproduction of a mammal.

A   A sperm nucleus joins with an egg nucleus.
B   The embryo develops in the amniotic sac.
C   The embryo becomes attached to the uterus wall.
D   A fertilised egg passes down the oviduct.
E   The young animal is born.

Arrange the stages into the correct order by writing the letters into the boxes.

1

(*b*) The diagrams show the human female and male reproductive systems.

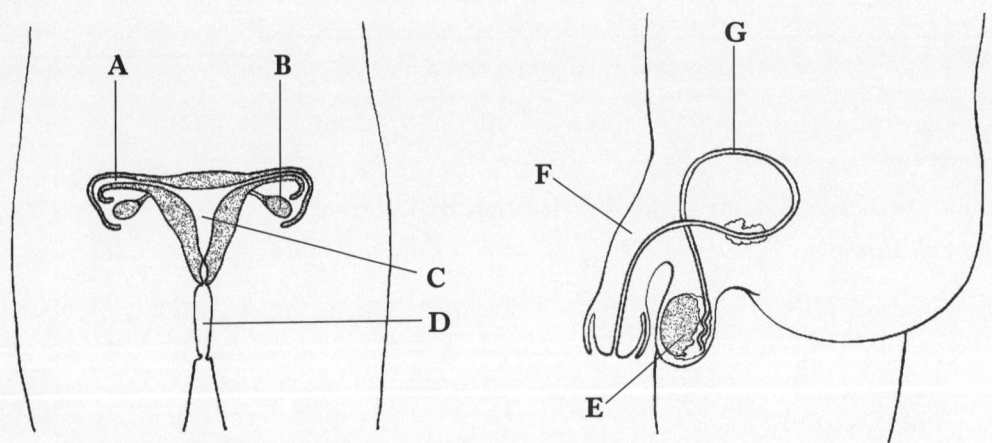

Complete the table below by adding the correct letter, name and function of the parts.

Letter	Name	Function
	ovary	
		where the embryo develops
E		

3

DO NOT
WRITE IN
THIS
MARGIN

Marks | KU | PS

7. (continued)

(*c*)   (i)   Mammal embryos obtain their food from their mother's blood.

Where do the embryos of fish obtain their food?

_____   1

(ii)   Young fish care for themselves.

How are young mammals cared for?

_____   1

[Turn over

8. (*a*) The diagram represents a section through the heart of a mammal.

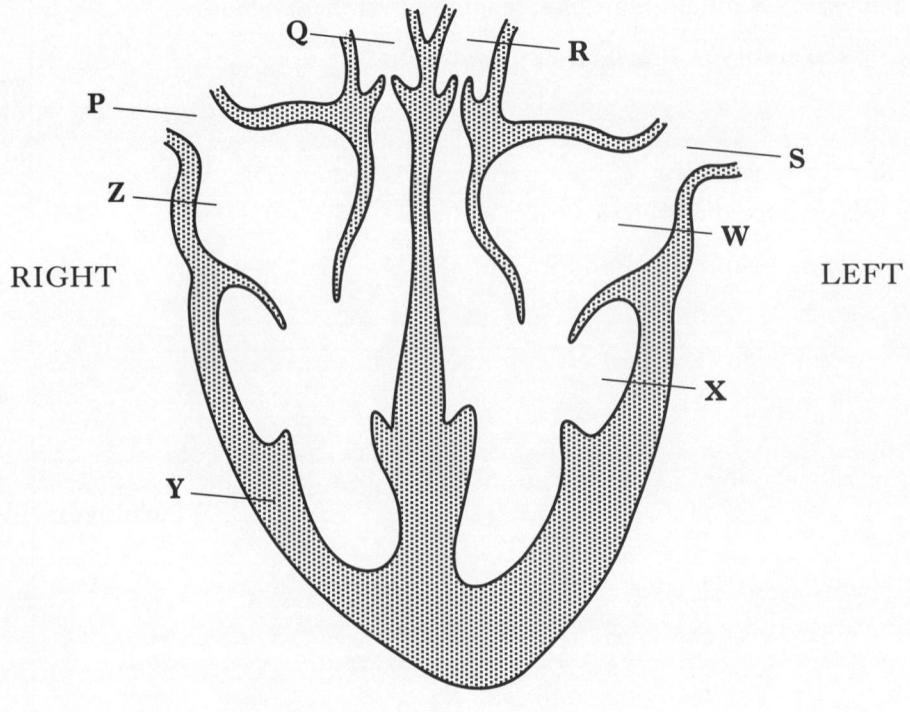

RIGHT   LEFT

Use letters from the diagram to identify the following.

   (i) The two atria (auricles).

      _____ and _____

   (ii) The vessel which brings blood to the heart from the lungs.

      _____

   (iii) The two vessels which carry deoxygenated blood.

      _____ and _____

(*b*) Explain why the wall of the left ventricle is thicker than the wall of the right ventricle.

_____

_____

1

1

1

1

DO NOT WRITE IN THIS MARGIN

Marks KU PS

8. (continued)

(*c*) (i) Use lines to connect each of the blood vessels to the correct description of blood flow.

Blood vessel	*Description of blood flow*
arteries	away from the heart
veins	through the tissues
capillaries	towards the heart

2

(ii) In which type of blood vessel may a pulse be felt?

__________________________ 1

[Turn over

Marks | KU | PS

9. The pulse rate of an athlete was monitored during a training exercise. The results are shown in the table.

Time (minutes)	0	1	2	3	4	5	6	7	8
Pulse rate (beats per minute)	65	65	90	118	118	118	105	100	80

(a) Complete the line graph of the results by

(i) labelling and adding an appropriate scale to the horizontal axis, **1**

(ii) plotting the graph. **1**

The first three points have been plotted.

(Additional graph paper, if required, will be found on page 31.)

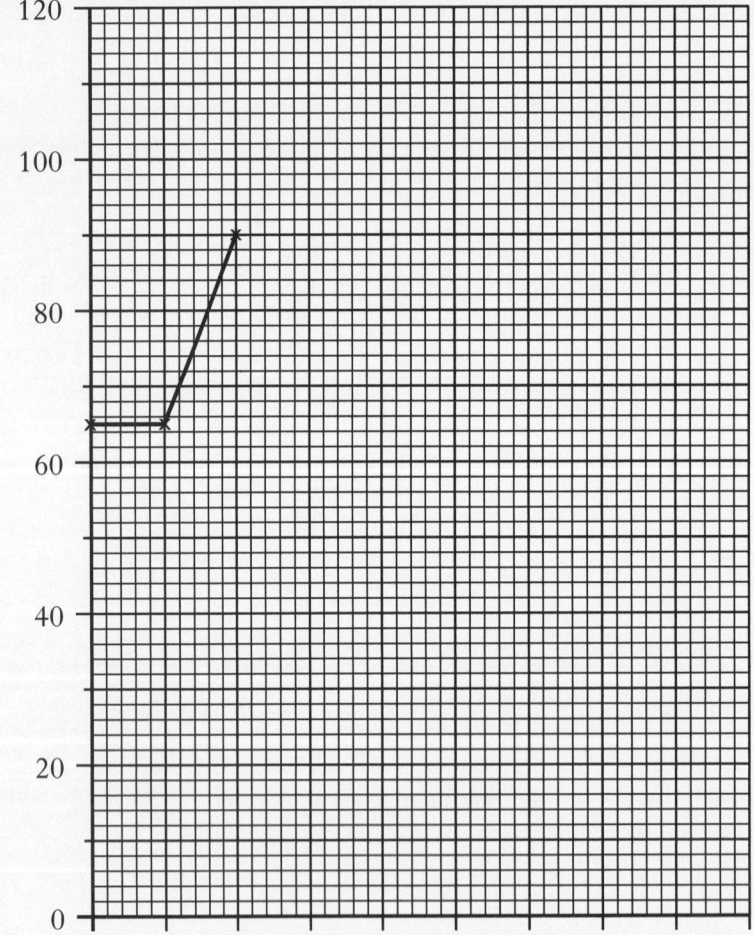

(b) What evidence suggests that the athlete had not fully recovered at 8 minutes?

_____ **1**

DO NOT WRITE IN THIS MARGIN

Marks	KU	PS

9. (continued)

(*c*) Suggest an improvement to the procedure to allow the recovery time of the athlete to be measured.

____

____ **1**

(*d*) How might the recovery time of the athlete differ from that of an untrained person?

____

____ **1**

[Turn over

Marks | KU | PS

10. Read the following passage and answer the questions based on it.

Hayfever

Hayfever affects 2 to 3 million people in Britain. It is caused by an allergy to pollen or sometimes the spores of fungi. The body's immune system reacts by releasing excess histamine. This results in an irritation and inflammation of the nose and eyes.

The symptoms vary and may involve sneezing, a runny or blocked nose, and a sore throat. The eyes may become red, watery or itchy. In addition, a wheezy chest may suggest that the sufferer also has asthma. The peak pollen time is early summer when school and university examinations take place. This can make it difficult to revise and perform well.

Hayfever is related to asthma and eczema. It is quite common to find members of the same family with one or more of these conditions.

Various treatments are available without prescription. These include antihistamine tablets to reduce the allergic response as well as nasal sprays and eye drops to reduce inflammation. For severe cases, doctors may prescribe either tablets or injections containing steroids. These can cause side effects so the benefits have to be weighed against the possible disadvantages. Tablets are more favoured than injections. Other types of injection can desensitise patients to the pollen causing their allergy. Unfortunately, they may produce serious side effects and, as they can only be given under close hospital supervision, are hardly ever used.

(*a*) What effect does pollen have on the body's immune system in hayfever sufferers?

_____ 1

(*b*) What evidence is there that hayfever might have a genetic component?

_____ 1

(*c*) Which symptom suggests that a hayfever sufferer may also have asthma?

_____ 1

DO NOT
WRITE IN
THIS
MARGIN

Marks	KU	PS

10. **(continued)**

(*d*) At what time of the year are hayfever sufferers likely to be worst affected?

_____

1

(*e*) What type of substance is found in treatments prescribed for severe cases of hayfever?

_____

1

(*f*) Why are desensitising injections not used very often?

_____

1

[Turn over

11. (*a*) The bar chart shows the results of a survey into blood groups of a sample of people in a Scottish town (Town X).

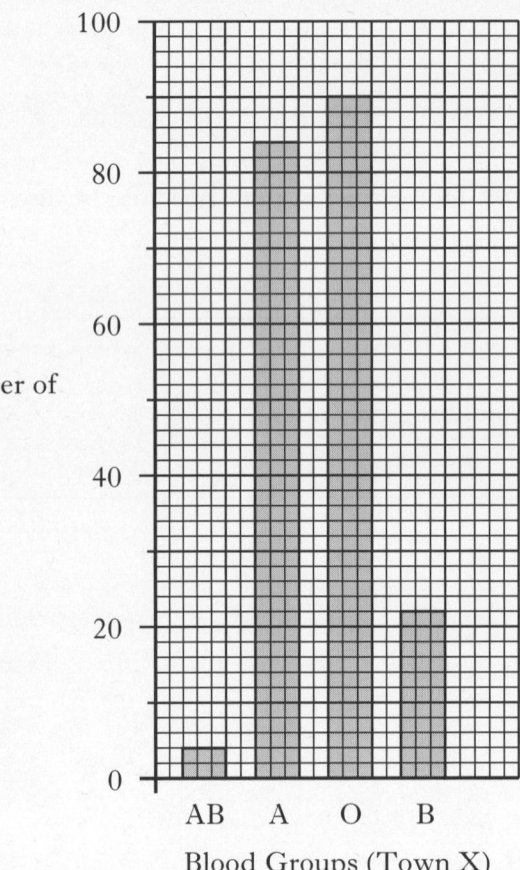

Number of people

Blood Groups (Town X)

(i) Is the variation in blood group continuous or discontinuous?

_____

1

(ii) How many people were in the survey?

Space for calculation

Number of people _____

1

Marks | KU | PS

11. (continued)

(b) A similar survey was carried out on a sample of 1000 people in a different town of the same size (Town Y).

There were 20 with group AB, 500 with group A, 400 with group O and 80 with group B.

(i) Complete the pie chart of these results by drawing and labelling the remaining segments.

(An additional chart, if required, will be found on page 31.)

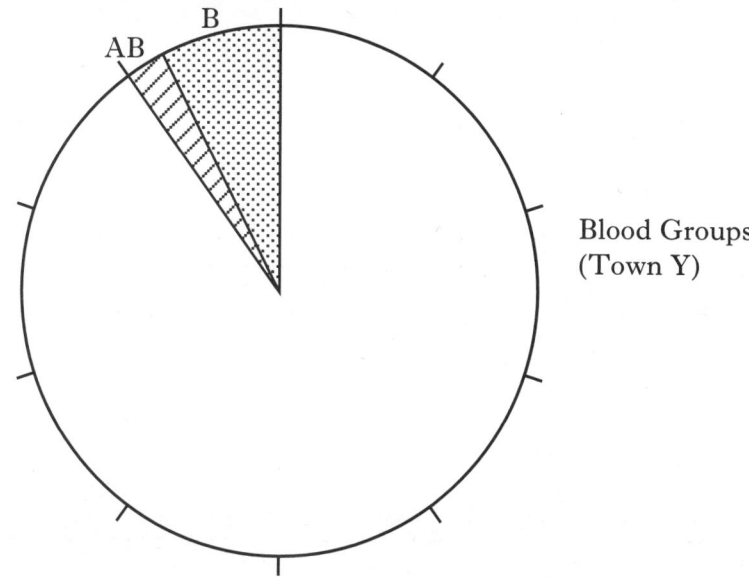

Blood Groups
(Town Y)

2

(ii) What percentage of people in the Town Y sample had blood group O?

Space for calculation

_____ %

1

(c) (i) Select **one** similarity and **one** difference between the results of the surveys for Towns X and Y.

Similarity _____

1

Difference _____

1

(ii) Which blood group survey, Town X or Town Y, is the more reliable?

Give a reason for your answer.

Town _____

1

Reason _____

1

[Turn over

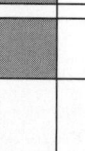

Marks | KU | PS

12. Maggots move away from light. The effect of different light intensities on the rate of movement was investigated using the apparatus shown below.

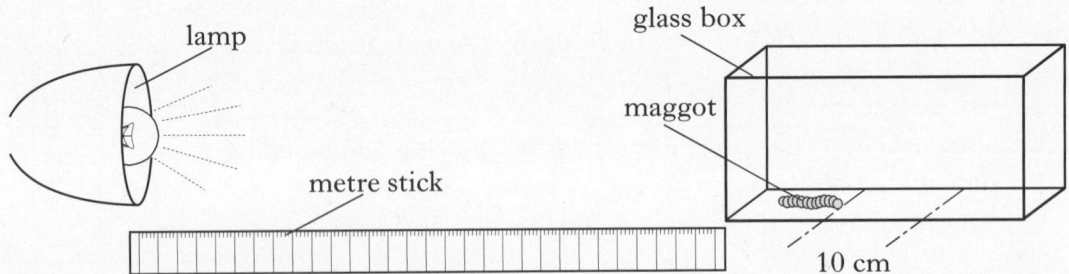

The time taken for a maggot to move 10 cm was recorded when the lamp was at different distances from the glass box. The experiment was carried out using three different maggots. The results are shown in the table.

Distance between lamp and box (cm)	Time taken to move 10 cm (seconds)			
	Maggot 1	Maggot 2	Maggot 3	Average
100	30	33	30	31
50	23	24	22	
25	18	20	19	19

(a) Complete the table with the average time for the maggots to move 10 cm when the lamp was at a distance of 50 cm.

Space for calculation

1

(b) Underline **one** option in each bracket to complete the following sentences correctly.

As the distance between the lamp and the glass box decreases, the light

intensity $\left\{ \begin{array}{l} \text{increases} \\ \text{decreases} \\ \text{stays the same} \end{array} \right\}$.

1

As the light intensity decreases, the time taken for the maggot to move

10 cm $\left\{ \begin{array}{l} \text{increases} \\ \text{decreases} \\ \text{stays the same} \end{array} \right\}$.

1

(c) What would happen to the rate of movement of the maggots if they were placed in darkness?

_____

1

13. Phosphorylase is an enzyme extracted from potatoes. Drops of phosphorylase, glucose-1-phosphate and water were added to a dimple tile as shown.

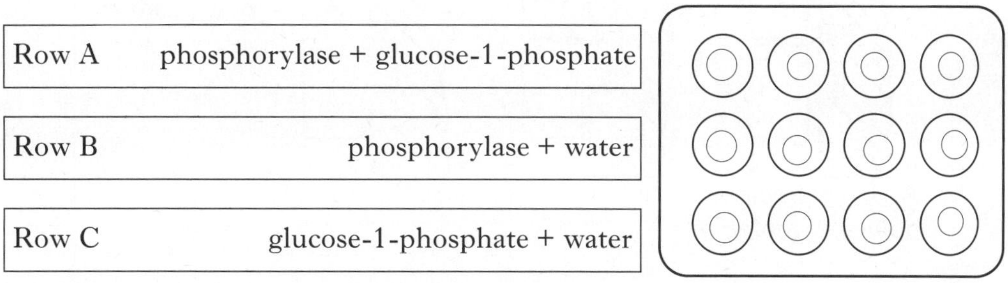

Row A	phosphorylase + glucose-1-phosphate
Row B	phosphorylase + water
Row C	glucose-1-phosphate + water

A drop of iodine solution was added to one dimple in each row at three-minute intervals. If starch is present, a black colour forms.

The results are shown below.

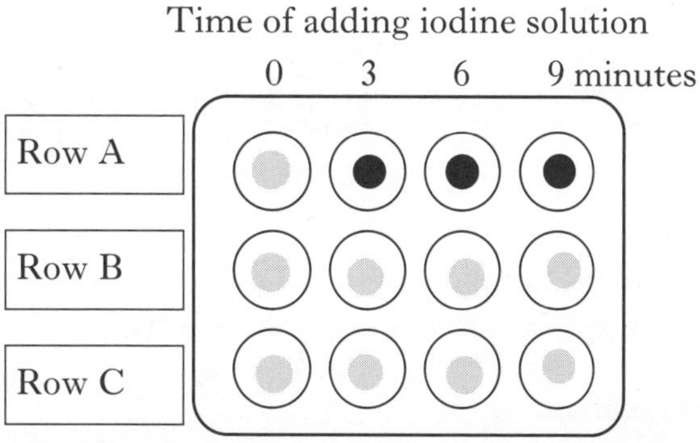

Time of adding iodine solution

0   3   6   9 minutes

Row A

Row B

Row C

(a) In which row has starch been synthesised?

Row _____

(b) The experiment was carried out at 25 °C. How would the results in Row A differ if the experiment had been carried out at a lower temperature?

_____

(c) Rows B and C are control experiments.

  (i) What conclusion can be drawn from Row B? _____

  _____

  (ii) What conclusion can be drawn from Row C? _____

  _____

[Turn over

Marks | KU | PS

14. The diagrams show two different types of enzyme-controlled reactions.

Diagram 1
Synthesis reaction

new product

enzyme   substrate
molecule   molecules

Diagram 2
Breakdown reaction

enzyme
molecule

2 new
products
formed

substrate
molecule

(a) For each of the following word equations state whether it is an example of a synthesis reaction or a breakdown reaction.

Word equation   *Type of reaction*

(i) maltose   **enzyme X**   glucose molecules _____

(ii) amino acid molecules   **enzyme Y**   protein molecule _____

(iii) fatty acids and glycerol **enzyme Z** fat molecule _____   2

(b) Of what type of substance are enzymes made?

    _____   1

(c) Respiration provides energy for cells to carry out various functions.

    Underline **two** of the following functions which require energy from respiration.

    Muscle contraction   Osmosis   Diffusion   Cell division   1

Marks | KU | PS

15. (*a*) The following grid contains some terms used in studying inheritance.

A gamete formation	B tallness in peas	C genotype
D phenotype	E true breeding	F gene
G fertilisation	H dominant	I dwarfness in peas

Use letters from the grid to identify the correct term for each of the following.

(i) Part of a chromosome _____ 1

(ii) Involves a reduction in the number of chromosomes _____ 1

(iii) Two different phenotypes of the same characteristic _____

and _____ 1

(iv) An organism with only one form of a particular gene _____ 1

(v) The genes that an organism contains for a characteristic _____ 1

(*b*) Pea plant cells contain 14 chromosomes.

(i) How many complete sets of chromosomes does this represent?

_____ sets 1

(ii) How many chromosomes are there in the sex cells of pea plants?
Space for calculation

_____ chromosomes 1

[Turn over

DO NOT
WRITE IN
THIS
MARGIN

Marks | KU | PS

16. The diagram shows the apparatus used to produce large numbers of bacterial cells for manufacturing insulin.

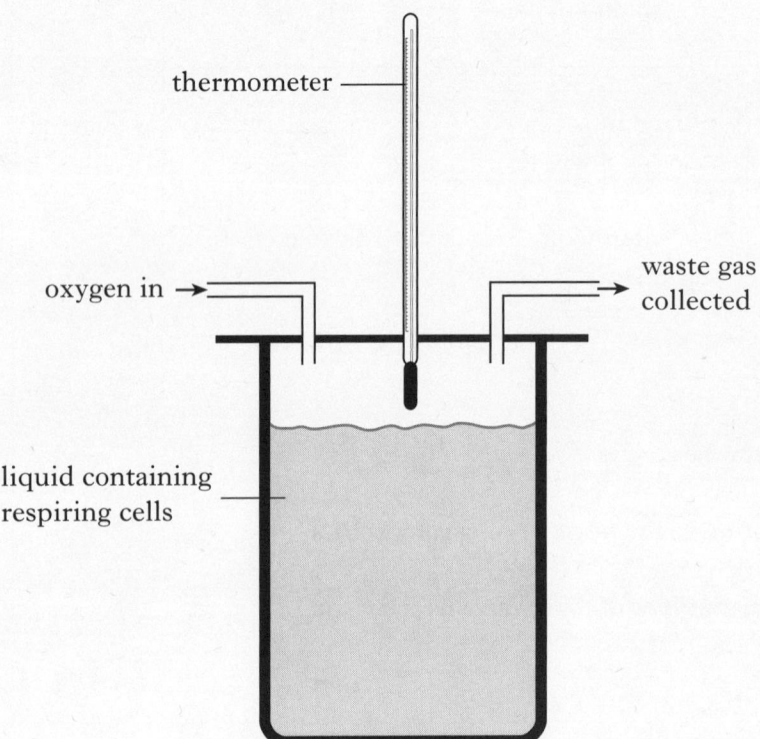

thermometer

oxygen in →

waste gas collected

liquid containing respiring cells

(a) Suggest an improvement which could be made to the way the apparatus is set up and explain why it is necessary.

Improvement _____   1

Explanation _____

_____   1

(b) (i) Which type of respiration will take place because of the presence of oxygen?

_____   1

(ii) What additional factor, not shown in the diagram, must be supplied to allow the bacteria to respire?

_____   1

DO NOT WRITE IN THIS MARGIN

	Marks	KU	PS

16. (*b*) (continued)

(iii) What waste gas will be produced during respiration?

____ **1**

(iv) What form of energy, other than chemical, may be released by the bacterial cells during respiration?

____ **1**

[Turn over

DO NOT WRITE IN THIS MARGIN

Marks | KU | PS

17. A sewage works removes organic material before water is discharged into a river. This is done in two main stages.

Stage 1 Organic solids settle out as sludge which is treated separately.

Stage 2 The remaining liquid is treated with living organisms.

(*a*) (i) Name **one** useful product which can be made from the treated sludge produced in Stage 1.

______________________________ **1**

(ii) What type of organisms act on the liquid in Stage 2?

______________________________ **1**

(iii) Describe **one** way in which oxygen can be provided for the organisms in Stage 2.

____

____ **1**

17. (continued)

(b) When water from a sewage works is analysed, several measurements are made. The table shows some of the measurements taken over one year.

Month	Suspended solids (mg/l)	Biochemical oxygen demand (mg/l)
January	35·0	31·0
February	42·0	40·0
March	44·0	35·5
April	30·5	18·0
May	27·0	17·0
June	29·5	19·0
July	21·5	14·5
August	25·5	16·5
September	25·5	16·5
October	29·5	22·0
November	34·5	28·5
December	32·5	35·0

(i) Sewage works should not discharge water with more than 30 mg/l suspended solids **and** a biochemical oxygen demand of more than 20 mg/l.

In which months of the year was water from the sewage works not meeting this standard?

_____

_____   1

(ii) Suggest **one** abiotic factor which affects how well the living organisms break down the sewage over the course of a year.

_____   1

[Turn over

Marks | KU | PS

18. An investigation was carried out into the effects of various additives on dough. Yeast was mixed with flour and sugar solution to make dough. The dough was then cut into four pieces and additives were added to three of them. 20 cm³ of each dough was put into measuring cylinders and the volume of the dough was measured after one hour.

The results are shown below.

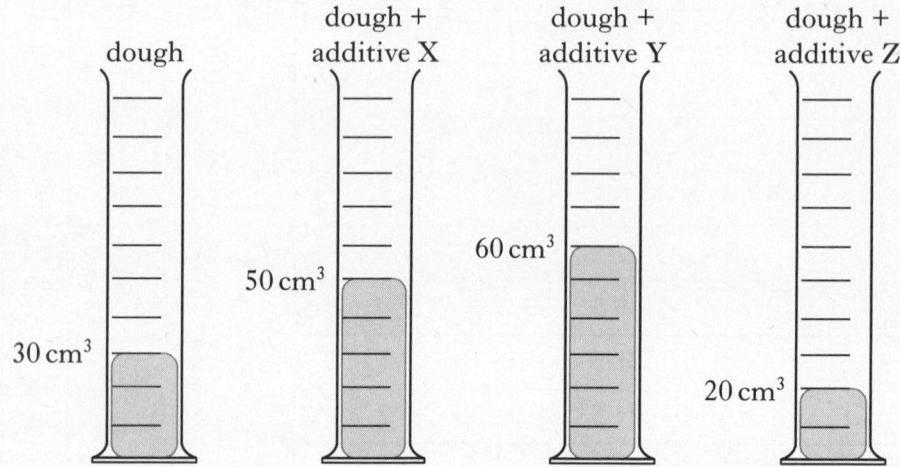

dough

dough +
additive X

dough +
additive Y

dough +
additive Z

60 cm³

50 cm³

30 cm³

20 cm³

(a) Calculate the percentage increase in the volume of the dough with no additive.

Space for calculation

_____%   1

(b) What substance produced by yeast caused the dough to rise?

_____   1

(c) Which additive caused the greatest increase in the volume of the dough?

_____   1

(d) Which additive may have prevented the yeast fermenting?

_____   1

(e) What type of organism is yeast?

_____   1

DO NOT WRITE IN THIS MARGIN

Marks KU PS

19. (*a*) Pollution can affect areas such as fresh water and seas.

Name the **two** other main areas which can be affected by pollution.

1 ______________________

2 ______________________ 1

(*b*) Complete the table to show the **three** main sources of pollution and **one** example of a pollutant from each.

Source	*Example of pollutant*
industry	
	fertilisers
	litter

2

(*c*) Pollution from the exhausts of vehicles can be a major problem in some cities.

Give **one** way in which this type of pollution can be controlled.

____

____ 1

[*END OF QUESTION PAPER*]

[Turn over

SPACE FOR ANSWERS
AND FOR ROUGH WORKING

ADDITIONAL GRID FOR QUESTION 2(*c*)

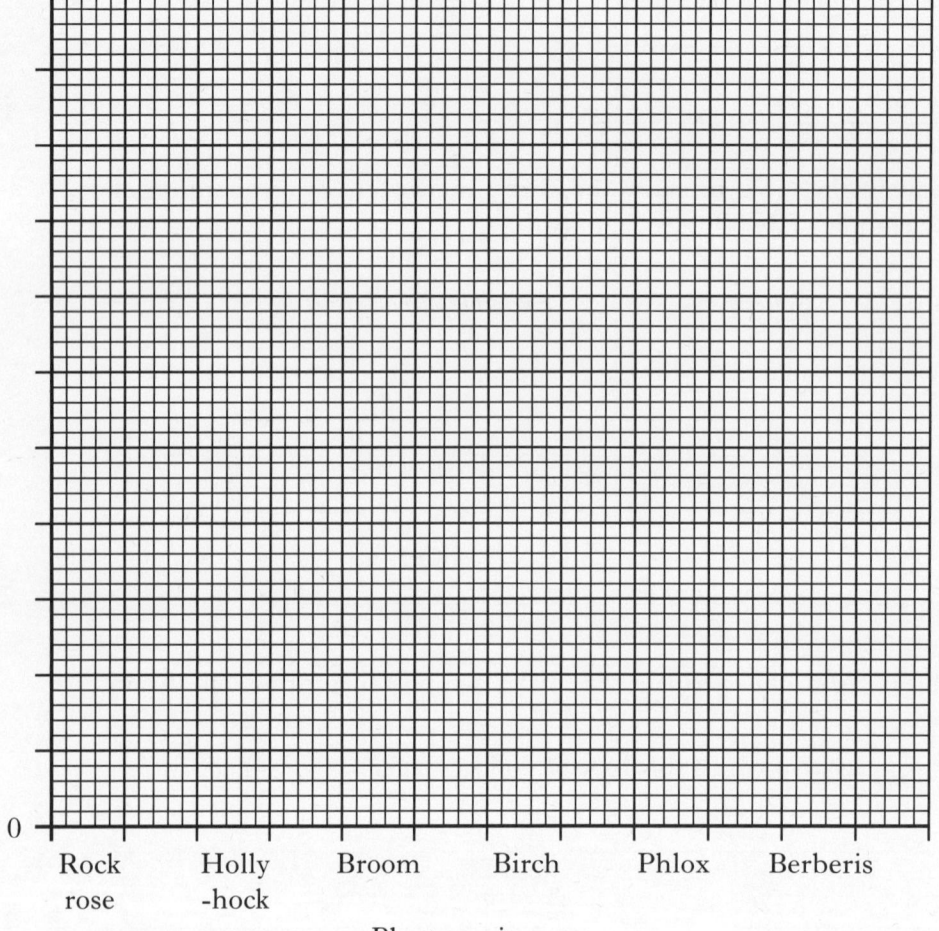

SPACE FOR ANSWERS
AND FOR ROUGH WORKING

ADDITIONAL GRID FOR QUESTION 9(a)

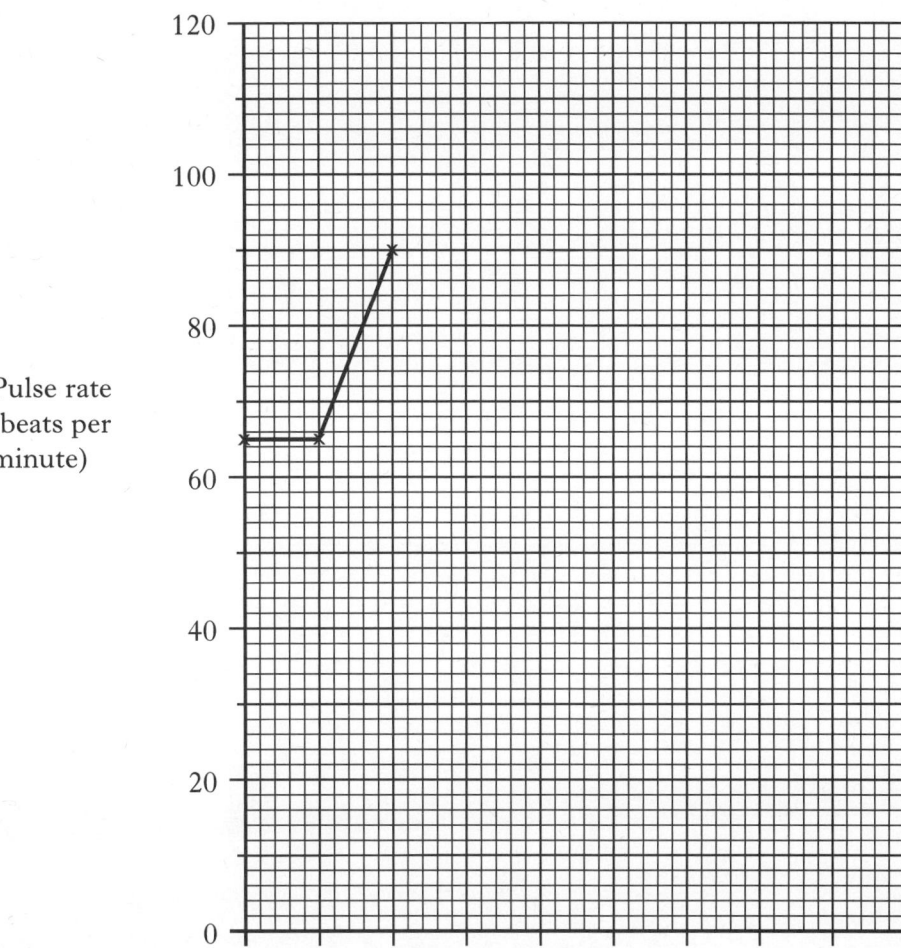

ADDITIONAL GRID FOR QUESTION 11(b)(i)

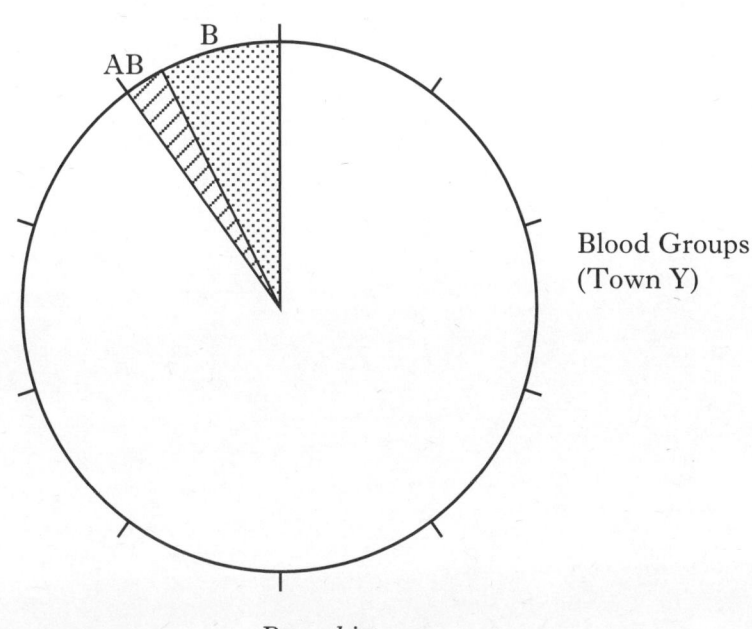

Blood Groups
(Town Y)

SPACE FOR ANSWERS
AND FOR ROUGH WORKING

[BLANK PAGE]

[BLANK PAGE]

[BLANK PAGE]

[BLANK PAGE]

[BLANK PAGE]

Acknowledgements

Leckie & Leckie is grateful to the copyright holders, as credited, for permission to use their material:
Phillip Allan Publishers Ltd for 'Designing a Sports Drink' from *Biological Sciences Review,* September 2002 (2002 paper p 14).

The following companies/individuals have very generously given permission to reproduce their copyright material free of charge:
The Sunday Herald for an extract from the article 'Salt Sellers Threaten The Whale' (2001 paper p 20);
The Dairy Council for an extract from *Yoghurt Factsheet* (2003 paper p 18);
The Herald for the article 'Stirring Stuff's in the Bag' (2004 paper p 16).

Pocket answer section for
SQA General Biology
2001 to 2005

Published by Leckie & Leckie Ltd, 8 Whitehill Terrace, St Andrews, Scotland, KY16 8RN
tel: 01334 475656, fax: 01334 477392, enquiries@leckieandleckie.co.uk, www.leckieandleckie.co.uk

Biology General Level 2001

1. (a) organisms that eat other organisms/ organisms that do not make their own food

(b) arrows

(c) movement (one example)/heat/ respiration/waste (one example)

Any two

2. (a)

Ecosystem affected	Source of pollution	Example of pollutant
Air	Domestic	CFC gases from aerosol sprays
Fresh water	**agriculture/forestry/ farming/use of pesticides on crops/ run-off from fields**	Pesticides in a river
seas/oceans/salt water/coasts/ shorelines	Industrial	Crude oil from tanker vessels
Land	Domestic	**litter/sewage/ rubbish**

2. (b) **X** use alternative fuel/energy source/ remove acidic gases from smoke/ scrubbing/filters

Y sewage treatment/discharge into sea

Z use alternative fuels/fit more efficient exhaust systems/use unleaded/low sulphur petrol/increased use of public transport/smaller cars/reducing traffic or example/ban inefficient cars

3. (a)

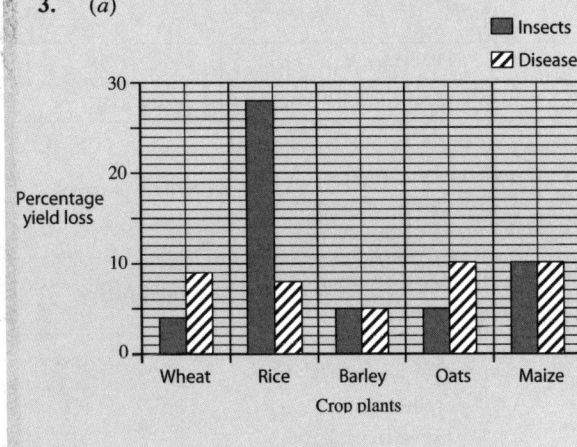

3. (b) barley

(c) wheat   oats

4. (a) (i) 20   (ii) 42   (iii) 21000

(b) food/raw materials/medicine/ fabrics/decorative/perfume/fuels/green fertiliser/preventing soil erosion
Any two

5. (a)

Letter	Name	Function
A	**embryo**	forms young plant
C	seed coat	**protection**
B	food store	resources for growth

(b) (i) 20

(ii) W Pansy   X Dahlia
    Y Marigold

6. (a) (i) carbon dioxide   (ii) stomata/stoma

(b) oxygen

(c) starch

(d)

Transport tissue	Substance carried	Part of plant from which substance is carried
xylem	**water/ minerals**	root
phloem	sugar	leaves

7. (a)

Letter	Cell part
A	**cell wall**
B	**chloroplasts**
C	**vacuole**

(b) mitosis/mitotic

(c) growth/repair/metabolism/chemical reactions/movement of substances

(d) Glucose + **oxygen** ⟶ **carbon dioxide** + **water** + energy

8. (*a*)　(i)

Letter	Name of part
A	**salivary glands**
B	oesophagus
D	**pancreas**
F	liver
I	**appendix**

　　　(ii)　absorbs water

　　　(iii)　1.　allows time for digestion/absorption/

　　　　　provides large surface area for absorption

　　　　　2.　allows rapid absorption/ transport of digested food

　(*b*)　(i)　The **renal artery** brings blood to the kidney and the **renal vein** takes blood away from the kidney.

　　　(ii)　The kidneys are the main organs for regulating the water content of mammals.

　　　　Their method of action involves **filtration** of blood followed by **reabsorption** of useful substances.

　　　(iii)　**urea** is a waste product which is removed in the **urine**

9. (*a*)　(i) (ii) (iii)

Tooth number	Starting/First weight (mg)	Final/Second weight (mg)	Loss in weight (mg)	Percentage loss in weight
1	3000	2100	900	**30**
2	4200	**3780**	**420**	10
3	3800	3040	760	20
4 (control)	4000	4000	0	0

　(*b*)　for comparison with other results/to prove single factor is the cause/

　　　to show it was the cola which caused loss of weight

　(*c*)　so bacteria did not affect results/to ensure they were safe to handle

　(*d*)　temperature/time/type of cola/ concentration of cola/type of teeth/age of teeth/state of decay

　(*e*)　cola causes a decrease in weight of teeth/ cola dissolves teeth

10. (*a*)　35

　(*b*)　increase in activity decrease in activity

　(*c*)　19 mm

11. (*a*)

Letter	Name	Function
A	**ear drum**	picks up vibrations in the air
B	bones of the middle ear	amplify vibrations and pass them to the inner ear
E	**cochlea**	changes vibrations into nerve impulses
D	auditory nerve	**pass nerve impulses to the brain**
C	semi-circular canals	**detect movement/position/ angle of head**

　(*b*)　direction of sound

　(*c*)　nerves　spinal cord　brain

12. (*a*)　(i)

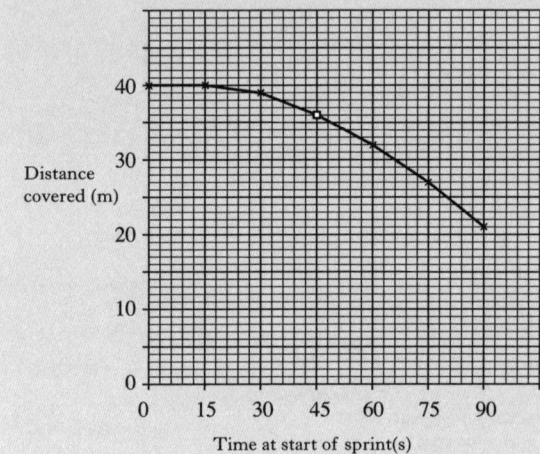

　　　(ii)　75 and 90

　　　(iii)　distance covered decreases

　　　(iv)　repeat with same athlete/repeat with other athletes

　(*b*)　Muscle fatigue is caused by the lack of **oxygen** and the build up of **lactic acid** in muscles.

13. (*a*)　(i)　seagulls　　(ii)　1100

　(*b*)　(i)　lapwings and pigeons

　　　(ii)　no effect　　(iii)　5 : 1

14. (*a*)　Yeast is a **fungus** and is **single-celled**.

　　　Yeast can use **sugar** as a source of food.

　(*b*)

Description	Word
Organisms used to make yoghurt	**bacteria**
Pieces of these can be transferred from a different organism into bacteria by genetic engineering to make new substances	**chromosomes**
Chemicals made by micro-organisms and which kill bacteria	**antibiotics**

15. (*a*)　seas around Alaska

　(*b*)　shelter buoyancy/support

15. (c)   seven million

(d)   glass
cosmetics

(e)   pumping salt-free water into lagoon

(f)   no
claimed due to chemical/dye released by
drug traffickers

16. (a)   (i)

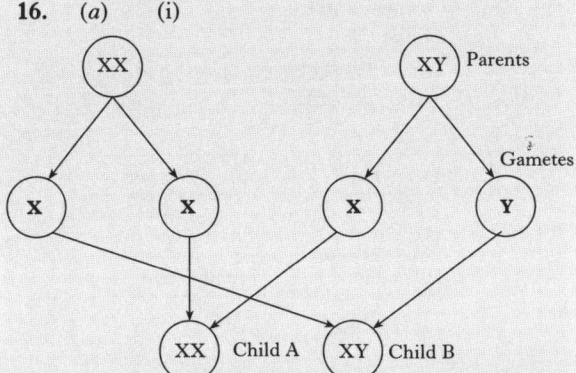

(ii) A–female   B–male   (iii) F1

(b)   (i) genotype   (ii) gamete
(iii) variations

17. (a)   respiration/fermentation in tube D
live yeast using sugar and producing
bubbles/CO_2

(b)   decrease/none

(c)   A + B only ☐   B + C only ☐

A + C only ☐   A, B + C ☑

18. (a)   Wild – high yield, poisonous fruits, strong
roots
Cultivated – low yield, good fruits, weak
roots

(b)   (i)   X1 and X4

(ii) both have good fruits and strong
roots/both have two good qualities

(c)   Y1 and Y4

(d)   selective breeding/artificial selection

19. (a)   (i)

Type of wash	Enzyme X (% activity)	Enzyme Y (% activity)
Warm (40 °C)	**59**	56
Medium (50 °C)	41	60
Hot (60 °C)	19	**50**

(ii) Y   (iii)   activity is reduced

19. (b)   Enzymes are found in **all** cells and are
made of **protein**.
Enzymes are **catalysts** and work best in
warm conditions.

20. (a)   microorganisms/bacteria/
decomposers/unicellular

(b)   methane/biogas – fuel
collected solids – fertiliser

Biology General Level 2002

1. (a) (i) 4   (ii) 1500

   (b) Take more samples/use more quadrats

   (c) (i) sunlight/moisture/pH/soil fertility/
   wind/temperature/humidity/carbon
   dioxide level

   any 2, both needed

   (ii)
light	light meter
moisture	moisture meter
pH	pH meter/pH paper/pH indicator
soil fertility	soil analysis
wind	wind gauge/anemometer
temperature	thermometer
humidity	hygrometer/wet and dry thermometer
carbon dioxide	carbon dioxide meter

2. (a) (i)
   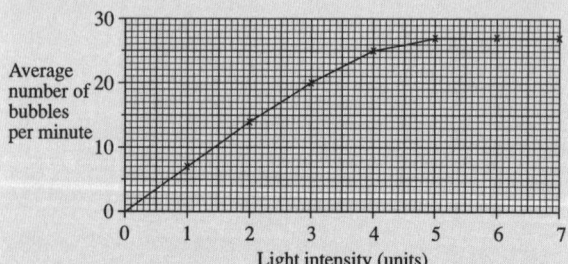

   (ii)
beetles	pit fall trap/pooter/beating branches
caterpillars	beating branches/sweep netting/pitfall trap
slugs	pitfall trap
woodmice	small mammal traps
spiders	beating branches/sweep netting/pitfall trap/pooter
chaffinches	netting
owls	netting
hawks	netting

   (iii) They feed on the same animal or
   They both eat woodmice

   (iv) The population of one may fall

   (b) habitat
   population
   community

3. (a)

   [Graph: Average number of bubbles per minute (y-axis, 0 to 30) vs Light intensity (units) (x-axis, 0 to 7). Line rises steeply then levels off around 27-28 from intensity 5-7.]

   (b) No effect/bubbling stayed the same

   (c) Use brighter bulbs/use dimmer bulbs/put
   filters in front of lamp/use dimmer
   switch/change the number of lamps/change
   distance between lamp and plant

3. (d) Increases reliability of results/reduces effect
   of atypical result

   (e) Oxygen

4. (a) (i) ✓ B   ✓ E

   (ii) water/moisture
   oxygen
   temperature
   light

   (b)
   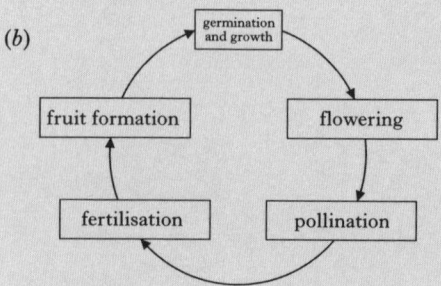

   (c) wind
   insects/named appropriate animal (eg bats/
   humming birds)

5. (a) (i) So no starch present/to remove starch

   (ii) Same except no chemical/replace
   chemical with inactive one

   (b) (i) starch   (ii) chlorophyll chemical

6. (a)
Source of Vitamin C	% of daily intake
Milk and dairy products	5
Fruit	35
Vegetables	45
Soft drinks	5
Other foods	10

   (b) (i) meat   (ii) vegetables

   (c) soft drinks

7. (a) A liver   B stomach   C pancreas

   (b) small intestine   (c) enzyme

   (d) growth or repair/maintain health/prevent
   disease/supply vitamins or minerals/provide
   roughage/water

8. (a)
Part of skeleton	Organ receiving protection
Skull	brain/eye/ear
Rib cage	brain/eye/ear
Backbone	spinal cord

   (b) tendons

   (c) Statement 1   minerals/calcium phosphate
   Statement 2   fibres/protein

   (d) (i) X   (ii) lactic acid

9. (a) glycogen

9. (b) reduction in energy store/glycogen/
carbohydrate or overheating or fluid loss/
dehydrating/reduced blood volume

(c) conserves energy store/reduces fatigue/delays
fatigue/supplies energy/replaces carbohydrate

(d) (i) increases heat loss/reduces overheating

(ii) increases water loss/dehydration/reduces
blood volume

(e) carbohydrate

water

sodium

(f) for taste

10. (a) rhythmical behaviour

(b) fry and parr are in fresh water and sea lice
only affect fish in sea

(c) yolk sac

(d) 400

(e) (i) July   (ii) 15   (iii) 25

11. (a) cells

stains

(b) **Q + R**

(c) (i) from . . .high concentration

to . . .   low concentration

(ii)

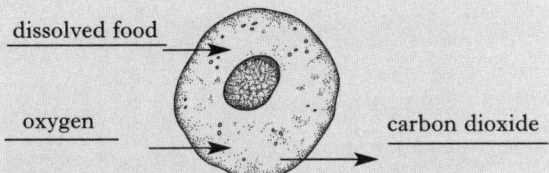

dissolved food

oxygen

carbon dioxide

(iii) membrane   (iv) osmosis

12. (a) (i) 41   (ii) it decreases

(iii) Effect   eggs would not hatch
Reason   embryos killed

(b) (i) speeds up chemical reactions without
being altered

(ii) Breakdown   amylase/pepsin/trypsin/
lipase/catalase/protease

Synthesis   phosphorylase

13. (a)

2		2
1		1
	2	

(b) gametes   (c) fertilisation

(d)

Y		X	X		X
XY/YX				XX	
		female		female	

14. (a) (i)

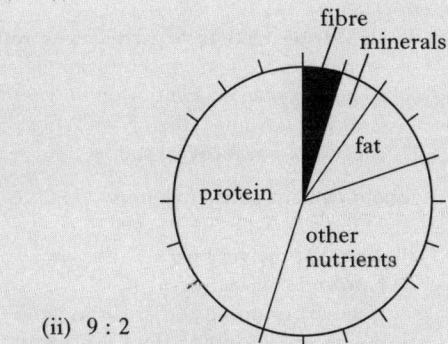

☐ Strathclyde
◩ Scottish average

Household waste recycled (%)

1990   1991   1992
Year

(ii) 1 : 2   (iii) Increasing

(b) (i) freshwater/sea water/water/land

(ii) cholera/typhoid/dysentery/polio

15. (a) (i) 200   (ii) A + B

(iii) Any **two** from:
lack of nutrient or food/lack of oxygen/
accumulation of wastes/lack of space/
change in pH

(b) (i)

fibre
minerals
fat
protein
other
nutrients

(ii) 9 : 2

(c) Asexual/binary fission

16. (a) (i) 30

(ii) 60

(b)

☐

☐

☑   20 – 30 min

☐

(c) repeat it

(d) (i) The same but without using yeast/The
same except using dead yeast

(ii) control

Biology General Level 2003

1. (a) Food web

   (b) (i) Producer – oak tree
   Consumer – woodlice/beetles/worms/
   squirrels/spiders/hedgehogs/blackbirds/
   foxes/hawks
   (ii) transfer of energy
   (iii) oak leaves → beetles → spiders →
   hedgehogs → foxes
   (iv) acorns
   (v) leaves

   (c) • habitat = where an organism lives
   • population = all the animals or plants of a
   single species living in an area
   • ecosystem = a particular area and all the
   animals and plants which live there

2. (a) 13 000

   (b) (i) A
   (ii) C = carbon monoxide
   D = smoke

3. (a) 2·8

   (b) 50

   (c) decreases (or equivalent) ... increases (or
   equivalent)
   (d) B smaller decrease in oxygen concentration/
   increase in oxygen concentration begins
   sooner/oxygen concentration downstream
   (from B) is higher (than from A)/oxygen
   concentration recovers sooner

   (e) typhoid/cholera/polio/dysentery

4. (a) (i) Found at the very back of the jaw   A
   Known as an incisor   C
   Used for grinding and crushing food   A
   (ii) killing prey/gripping (food)/(tearing) food

   (b) (i) 3
   (ii) 4
   (iii) 1. greater protection/more effective/fewer
   decayed teeth
   2. same protection with lower
   concentration/fewer side effects/less
   expense/avoids using excess fluoride

5. (a) (i) breath/breathing/exhaling
   (ii) 150
   (iii) 20

   (b) kidneys

   (c) urea

---

5. **(continued)**

   (d)
   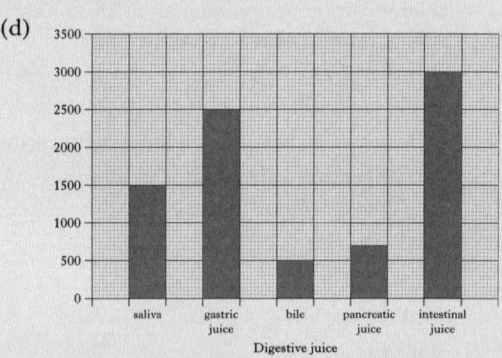

   (e) large intestine/colon

6. (a) (i) stomata/stoma
   (ii) carbon dioxide
   (iii) water

   (b) starch

   (c) chlorophyll

7. (a) (i) B and C
   (ii) A

   (b) B graft/grafting
   D tubers

8. (a)
   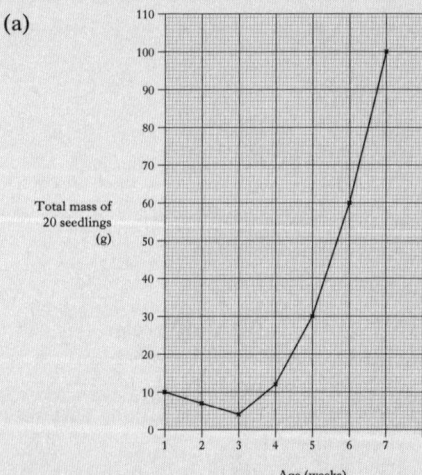

   (b) (i) 3
   (ii) 6–7 weeks
   (iii) 5

   (c) (i) Any two from:
   temperature/light/water/nutrients/type of
   soil/type of container/soil pH/humidity/
   planting depth/seed spacing
   (ii) increases reliability/reduces effect of
   individual variations/to make sample or
   results representative
   (iii) removing excess water/drying

   (d) no increase in mass/continued loss in mass

9. (a) cytoplasm   ✓
   nucleus   ✓
   cell membrane   ✓

9. (b) make them easier to see/increase visibility, clarity, contrast/to make parts of the cells easier to see/to make parts of the cells visible because cells are transparent

   (c) 120

10. (a) W glucose
       X oxygen
       Y water

   (b) food

   (c) Any two from:
       heat/growth OR
       repair/transport/movement/cell division OR mitosis

11. (a) (i) 44
       (ii) 6

   (b) Water molecules moved into the funnel ✓

   (c) any value in range 84–90

12. (a) Middle East yoghurt is more acidic and thinner

   (b) skimmed (milk)/evaporated (milk)/dried (milk)

   (c) kill bacteria/reduce growth OR activity of bacteria

   (d) lactic (acid)

   (e) heating (to 85–95°C)/Pasteurisation

   (f) slower bacteria growth/fermentation/production of lactic acid
       slower bacterial activity

   (g) increase in acidity

13. (a) (i) 1. gold (body)   2. black (body)
       (ii) black (body)
       (iii) 4:1

   (b) All the F_1 generation have the same genotypes and phenotype ✓

   (c) discontinuous

   (d) gametes

14. (a) movement/muscle attachment/support/makes blood cells/framework for body

   (b) (i) Ligament
       (ii) protects bones/cushions bones/shock absorber/reduces friction/allows smooth movement/stops bones rubbing together

   (c)

Range	Type	Example
one plane	hinge	knee/elbow/finger/toe
many planes	ball and socket	hip/shoulder

15. (a) (i) B
       (ii) for comparison/to show that one factor is causing an effect/to show there is no carbon dioxide in unbreathed air/to show that the breathed air contains the carbon dioxide

15. (b) Any two from:
       • volume/amount of air breathed OR rate of breathing or number of breaths/how long air is breathed in and out
       • volume/amount of indicator
       • concentration of indicator/
       • colour OR pH of indicator
       • temperature
       • pupil

16. (a) The chamber that receives blood
       from the body   A
       The artery that carries blood from
       the heart to the body   F
       The chamber that pumps blood
       to the lungs   C
       The vein that carries blood from
       the lungs to the heart   G

   (b) red blood cells
       plasma

17. (a) genetic engineering OR genetic modification OR reprogramming microbes

   (b) insulin – treat diabetes/
       growth hormone – treat growth problems/
       Factor VIII – treating haemophilia/
       interferon – cancer treatment

   (c) asexual/binary fission

18. (a) pneumonia and skin abscesses

   (b) grow as single cells, spiral shaped cells, cause Lyme's disease

   (c) E. coli

Biology General Level 2004

1. (a) (i)

Consumer	Diet
Moths	**ferns, trees**
Ground-living insects	grass
Voles	**ferns, tree-living insects**
Weasels	mice, voles
Tree-living insects	trees
Shrews	tree-living insects, mites
Fungi	trees
Mites	**fungi**
Spiders	ground-living insects
Owls	**mice, voles, shrews**

(ii) ferns → moths → mice → weasels / owls

(b) Producer - something which makes food / carries out photosynthesis / gets its energy from the sun.

Consumer – something which eats / feeds off / gets its energy from other organisms / organic matter

2. (a)

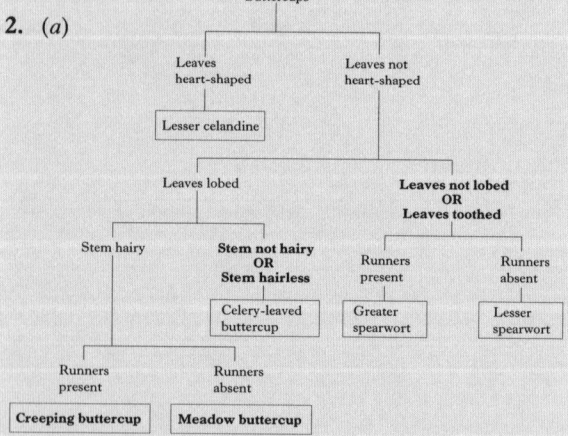

(b) leaf shape / one has heart-shaped leaves / one has toothed leaves

lesser celendine has heart-shaped leaves / lesser spearwort has toothed leaves

(c) leaf shape **and** absence of runners / both have lobed leaves **and** no runners

3. (a) (i)

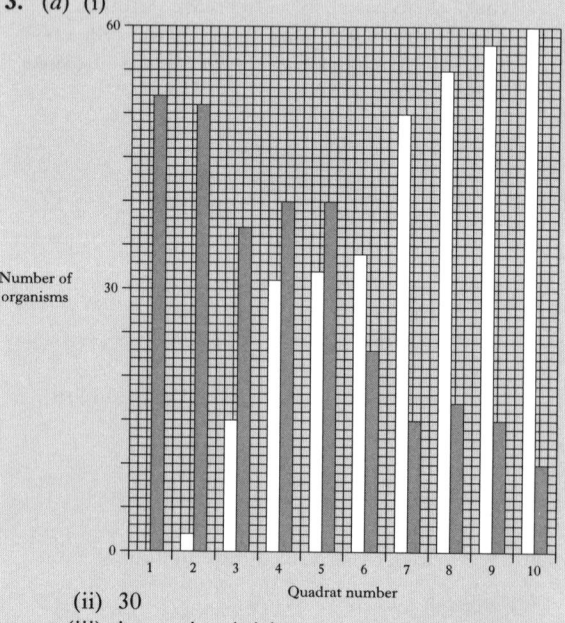

(ii) 30

(iii) increasing / rising

(b) increase in numbers / increased population

(c) water temperature, salt concentration

(d) habitat / abiotic factors

4. (a) (i) B
     carbon dioxide has not been removed

(ii) oxygen

(iii) so new starch could be detected / existing starch would prevent new starch being detected / so any starch found must have been made during the investigation / to see which plant can now make starch

(iv) species / type / size / age / health / leaf area / leaf number

(b) stoma / stomata / stomal pores

(c) chlorophyll

(d) (i) **A**   sepal

(ii) **E**   stigma

(iii) **F**   ovary

5. (a) (i) **J**   oviduct / fallopian tube
        K   uterus / womb
        L   ovary

(ii) Eggs produced   **L**
     Fertilisation takes place   **J**
     Fertilised egg becomes attached   **K**

5. (b)

Statement	Eggs	Sperm
Contain a food store for developing fetus	✓	
Swim using a tail		✓
Produced in testes		✓
In most fish, are deposited into the water	✓	✓
Are grametes	✓	✓

6. (a) Humidity / moisture (in the air) / water vapour

(b) They settle in the humid / moist area OR
They move to the humid / moist area OR
They move away from the dry area OR
They move to the side with water.

(c) To allow them to settle / to allow them to
explore conditions / to allow them to adjust to
conditions / to allow them time to move.

(d) To get a reliable / representative result
One may have been unreliable /
unrepresentative / atypical

(e) Light / temperature

(f) 1. Make both sides the same humidity / put
water in both sides / put drying agent in
both sides / make the bottom the same on
both sides / take out the water and the
drying agent.
2. Cover / shade one side OR shine light on
one side

7. (a)

Leaf cell	Cheek cell
A B C D E	A D E

(b) (i) Iodine solution + acetic orcein
(ii) Turns cytoplasm pink
(iii) Acetic orcein

(c) Stain

(d)

Power	Eyepiece lens magnification	Objective lens magnification	Total magnification
Low	× 12	× 4	× 48
Medium	× 12	× 10	× 120
High	× 12	× 40	× 480

8. (a) (i) 2
(ii) Diffusion

8. (b) (i) Water
(ii) (+) 20
(iii) 0.4
(iv) 0.25

9. (a) (about) 135 million cups

(b) Relaxation

(c) Heart disease

(d) Lung, bowel, digestive system

(e) Tannin, caffeine

(f) Any three of: Organic, Chai spice,
decaffeinated, herbal, iced

10. (a) 102

(b) Colour

(c) 20

11. (a) (i) It doubles
(ii)

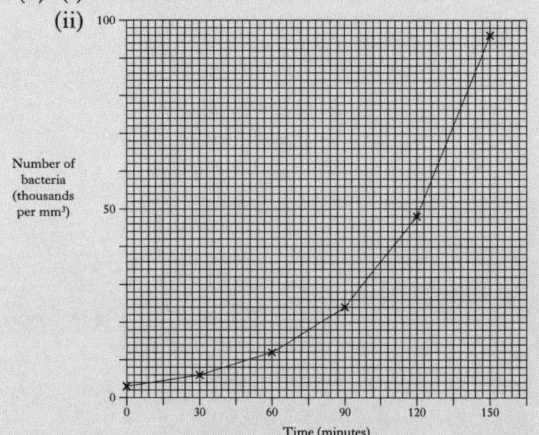

(iii) 768

(b) 1st stage **C**
2nd stage **B**
3rd stage **D**
4th stage **A**

(c) In comparison with the original cell, the
number of chromosomes present in a cell
produced by mitosis is **the same** and it
contains **the same** information.

12. (a)

Letter	Name of structure	Function
A	**Cornea**	Allows light to enter the eye
B	**Lens**	**Focuses light/ Produces a clear image**
C	Iris	**Controls amount of light entering / size of pupil**
D	**Retina**	Converts light into electrical impulses
E	Optic nerve	**Carries nerve signals / impulses / information**

(b) **Sight** –judgement of distance / 3D vision
Hearing –judgement of direction of sound

Biology General Level 2004 (cont.)

12. (*c*) (i) Muscles / effectors / glands
     (ii) 1 Brain
          2 Spinal cord

13. (*a*) (i) (Bone) marrow and lymph nodes
     (ii) White (blood cells)
     (iii) Red (blood cells)
     (iv) Protein
     (v) Engulf bacteria and produce antibodies
     (vi) 7200 million

  (*b*) (i) Capillary
     (ii) **H** – anywhere in blood vessel
          L – anywhere in muscle cells

  (*c*) Red blood cell / RBC / haemoglobin / oxyhaemoglobin

14. (*a*) They are different species

  (*b*) (i) Red

     (ii)

Generation	Symbol
A	P
B	F_1
C	F_2

     (iii) 4:1

15. (*a*) (i) Both have the thalassaemic gene
     (ii) Parents have affected children

  (*b*) 50% / half / 1 in 2 / 2 out of 4 / 1 to 1

  (*c*) Amniocentesis

16. (*a*) glucose / sugar → alcohol / ethanol + carbon dioxide + energy

  (*b*) 1. Making alcohol
      2. Baking

  (*c*) Yeast is a fungus and is single-celled.

  (*d*) 1. Wash with disinfectant / wash with alcohol/ cover with sterile material
      2. Flame / heat it till its red hot / put it in a bunsen

  (*e*) Prevent contamination (with unwanted microbes) / prevent escape of microbes / prevent escape of spores

Biology General Level 2005

1. (*a*) (i) movement/flow/transfer of energy
     (ii) 3
     (iii)

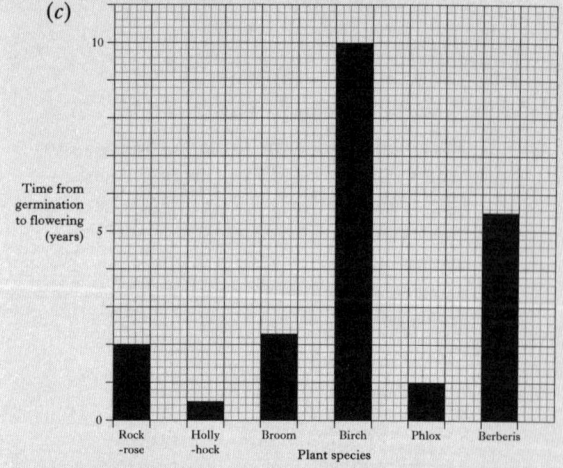

  (*b*) (i) woodlice numbers increase / more woodlice + less food for beetles / increased competition
     (ii) birth rate greater than death rate / death rate less than birth rate
     (iii) increase / increase in birth rate / decrease in death rate

  (*c*) micro-organisms
      nutrients
      plants

2. (*a*) 1

  (*b*) (i) A   embryo
         B   food store
     (ii) seed coat

  (*c*)

3. (*a*) 3376   4459

  (*b*) 278

  (*c*) 65

4. (*a*) chlorophyll

  (*b*) phloem

  (*c*) carbon dioxide and water

  (*d*) starch

5. (*a*) cover / shade one side

  (*b*) (i) humidity / moisture / temperature
     (ii) humidity / moisture – Put moist cotton wool / water or equivalent in base of one side. Put equivalent dry material / drying agent / nothing in other side

5. (*b*)(ii) (continued)

temperature - Surround one side with ice pack or equivalent. Surround other side with warm material.

6. (*a*) to make the results reliable / representative reduce effect of atypical result

(*b*) 144

(*c*) reaction time decreases

7. (*a*) A→D→C→B→E

(*b*)

Letter	Name	Function
B	ovary	produces / stores / releases eggs or female sex cells
C	uterus / womb	where the embryo develops
E	testis / testes / testicle	produces / releases sperm or male sex cells

(*c*) (i) yolk sac / egg yolk
(ii) by parents / adults

8. (*a*) (i) W   Z
(ii) S
(iii) P   Q

(*b*) left ventricle pumps blood further

(*c*) (i) arteries ——— away from the heart

veins ⟍ ⟋ through the tissues

capillaries ⟋ ⟍ towards the heart

(ii) arteries

9. (*a*)

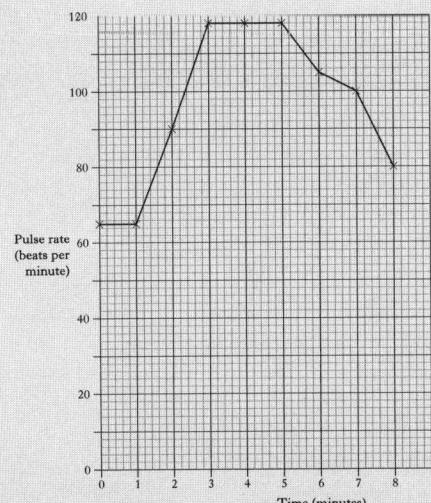

(*b*) pulse rate higher than start/pulse rate had not returned to normal/pulse rate still above 65

(*c*) continue timing until pulse rate returns to normal/65/ starting rate

(*d*) shorter/faster

10. (*a*) it releases excess histamine

(*b*) common for members of the same family to be affected / to have hayfever

(*c*) wheezy chest

(*d*) early summer

(*e*) steroids

(*f*) can cause serious side effects / they can only be given under close hospital supervision

11. (*a*) (i) discontinuous
(ii) 200

(*b*) (i)

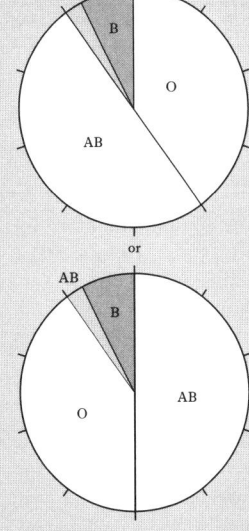

(ii) 40

(*c*) (i) Similarity - group AB lowest (in both) / group B 3rd (in both)
Difference - group O largest in X, not in Y / group A largest in Y, not in X
(ii) Y
larger sample size

12. (*a*) 23

(*b*) increases
increases

(*c*) decrease / moves slower

13. (*a*) A

(*b*) Longer for starch to be produced / less starch produced / longer to change colour

(*c*) (i) phosphorylase / enzyme alone will not produce starch / phosphorylase + G-1-P needed for starch production
(ii) glucose –1-phosphate alone will not produce starch / phosphorylase + G-1-P needed for starch production

14. (*a*) (i) breakdown
(ii) synthesis
(iii) synthesis

Biology General Level 2005 (cont.)

14. (*b*)   protein

(*c*)   muscle contraction   cell division

15. (*a*)   (i) F
(ii) A
(iii) B+I
(iv) E
(v) C

(*b*)   (i) 2
(ii) 7

16. (*a*)

Improvement	Explanation
oxygen inlet below surface	increased oxygen availability to bacteria / reduce waste of oxygen
thermometer bulb below surface	improved temperature measurement / to get temperature of the liquid
cooling jacket/ heating mechanism/ insulation	maintain (optimum) temperature
stirrer	spread oxygen / heat / nutrients / cells

(*b*)   (i) aerobic
(ii) food / nutrients / glucose / heat
(iii) carbon dioxide
(iv) heat

17. (*a*)   (i) methane / biogas / fuel / fertiliser
(ii) micro-organisms / bacteria / fungi / protozoa / microbes
(iii) (biological) filtration / filter bed / trickle / spray sewage over stones / air spaces in filter beds
activated sludge process / bubble air through sewage / stirring sewage

(*b*)   (i) January, February, March, November, December
(ii) temperature

18. (*a*)   50

(*b*)   carbon dioxide

(*c*)   Y

(*d*)   Z

(*e*)   fungus

19. (*a*)   air / atmosphere + land / soil / earth

(*b*)

Source	Example of pollutant
Industry	smoke / SO_2 / chemical waste / radioactive waste / oil / CO_2 / fumes / oxides of nitrogen / soot / nuclear waste / CO / noise / hot water / heavy metals /greenhouse gasses
agriculture / farming	fertilisers
domestic / people / household	litter

(*c*)   improved exhaust systems (or eg) / lead free petrol / traffic control measures (or eg) / increased use of public transport / electric vehicles / less harmful petrol/less use of cars/increase road tax